Capillarity and Wetting Phenomena

Pierre-Gilles de Gennes
Françoise Brochard-Wyart
David Quéré

Capillarity and Wetting Phenomena

Drops, Bubbles, Pearls, Waves

Translated by Axel Reisinger

With 177 Figures

 Springer

Pierre-Gilles de Gennes
Collège de France & Institute Curie
F-75005 Paris
France

Françoise Brochard-Wyart
Institute Curie
F-75005 Paris
France

David Quéré
Collège de France
F-75005 Paris
France

Cover Illustration: Pete Turner/Getty Images, 2002.

Library of Congress Cataloging-in-Publication Data

Pierre-Gilles de Gennes.
    Capillarity and wetting phenomena: drops, bubbles, pearls, waves/Pierre-Gilles
de Gennes, Françoise Brochard-Wyart, David Quéré.
        p. cm.
    Includes bibliographical references and index.

    1. Capillarity. I. Quéré, David. II. Gennes, Pierre-Gilles de. III. Title.

QC183.B874 2003
541.3'3—dc21   2003042431

This book is a translation of the French edition.

ISBN 978-1-4419-1833-8   e-ISBN 978-0-387-21656-0

Printed in the United States of America.   (EB)

9 8 7 6 5 4 3 2

springeronline.com

Preface

As I glance out my window in the early morning, I can see beads of droplets gracing a spider web. The film of dew that has settled on the threads is unstable and breaks up spontaneously into droplets. This phenomenon has implications for the treatment of textile fibers (the process known as "oiling"), glass, and carbon. It is no less important when applying mascara!

I take my morning shower. The moment I step out, I dry off by way of evaporation (which makes me feel cold) and by dewetting (the process by which dry areas form spontaneously and expand on my skin).

As I rush into my car under a pelting rain, my attention is caught by small drops stuck on my windshield. I also notice larger drops rolling down and others larger still that, like snails, leave behind them a trail of water. I ask myself what the difference is between these rolling drops and grains of sand tumbling down an incline. I wonder why the smallest drops remain stuck. The answers to such questions do help car manufacturers treat the surface of glass and adjust the tilt of windshields.

The traffic light suddenly turns red. I slam on the brakes and the car skids before finally coming to a halt. A firm grip on the road hinges on eliminating the film of water between tires and pavement. The car will stop only if direct contact can be established between the rubber and the asphalt, all in a matter of a few milliseconds.

The rain finally stops and I hear the squeaking sound of the windshield wipers rubbing against the glass. Friction between the rubber and the dry glass now opposes the movement of the wipers. Clever treatments of the glass can minimize that friction.

The sun is now shining and I hurry back to my garden to spray a fungicide onto a cluster of leaves covered with mildew. Unfortunately, drops fall off like so many beads, and only a small fraction of the product remains in place to perform its intended function. Is there a way to prevent the fungicide film from dewetting? Conversely, can one treat concrete (or the stones of historic monuments) to prevent them from soaking up water every time it rains?

These few examples illustrate the need to understand and tame the phenomenon of wetting. How can one turn a hydrophilic surface into one that is hydrophobic, and vice versa? We will describe a few solutions. Some rely on chemical treatments, such as coating a surface with a molecular layer of the right material. Others are rooted in physics, for instance, altering the surface roughness. We will also examine the dynamics of the wetting process. Drops spread spontaneously at a rate that slows with time. It may take years for a small drop to form a thin film covering a large surface area. In practice, films can be tricked by forcing them to spread suddenly. We will describe a few of their many-faceted dynamical properties.

When the word *bubble* is mentioned, most of us think of soap bubbles. Special additives are required for water to foam. The reason that a soap film can be made to stretch is just now beginning to be understood. Foams are desirable in a shampoo but can be a nuisance in a dishwasher detergent. Antifoam agents have been developed and have become commonplace, but how do they work? It is also possible to generate bubbles and foams without the help of surfactants, for example, in very viscous liquids such as glycerin, molten glass, and polymers. As we will see, the laws governing draining and bursting then turn out to be quite different from the conventional ones.

A child tosses a stone into a lake. He delights in watching capillary waves propagate by forming circular ripples on the water's surface. All of us have heard the sonic boom produced by an aircraft crossing the sound barrier. But how many of us are aware that we can also observe shock waves of capillary origin every day when we turn our kitchen faucet on: on the bottom of the sink water flows outward as a thin film. But a few centimeters away from the center, we see a hydraulic jump—very similar to a shock!

Our hope is that this book will enable the reader to understand in simple terms such mundane questions affecting our daily lives—questions that have often come to the fore during our many interactions with industry. Our methodology will consist in simplifying systems that often prove quite complex so as to isolate and study a particular physical phenomenon. In the course of developing models, detailed descriptions requiring advanced numerical techniques will often be replaced by an "impressionistic" approach based on more qualitative arguments. This strategy may at times sacrifice scientific rigor, but it makes it possible to grasp things more clearly and to dream up novel situations. Such is the spirit in which we wrote this book.

Paris, France   Pierre-Gilles de Gennes
                                                  Françoise Brochard-Wyart
                                                             David Quéré

Contents

Introduction

Years ago, Henri Bouasse wrote a classic French text on the topic of capillarity.[1] Bouasse has long been something of a celebrity in his field, not just on the strength of his technical writings but also because of his biting prefaces, in which he excoriated some of his colleagues. He was particularly intolerant of the professors at the Collège de France, who were not burdened by heavy teaching loads and devoted much of their time to such esoteric topics as the then nascent quantum physics. Bouasse failed to understand the physics revolution of the 20th century; yet he contributed with his legendary flair a number of enduring advances in classical physics, notably in his book on surface phenomena.

Eighty years later, capillarity continues to be a science in development. The Russian school led by Derjaguin worked on capillarity problems for 50 years.[2] In 1959, Mysels, Shinoda, and Frankel published their famous text on soap films.[3] Zisman, motivated by applied research on the lubrication of the clockwork of timepieces, elucidated the criteria for wetting.[4] Tanner, an aeronautics engineer, and Hoffman, a chemist, determined the experimental laws governing the spreading of liquids.[5] Many new concepts have emerged. Hence our incentive to write a new book.

We wanted to do it in the Bouasse tradition, that is to say, by aiming at an audience of students. What we offer here is not a comprehensive account of the latest research but rather a compendium of principles. Also following in Bouasse's footsteps, we do not claim to provide a detailed, up-to-date bibliography. All through these chapters, we suggest but a few major references with little regard for historical chronology.

We have endeavored to maintain as simple a presentation as possible. Our treatment is even less mathematical than was Bouasse's with its cycloids and other analytical tricks. Our goal is to illustrate concepts rather than to delve into detailed quantitative derivations. Even within this framework, we had to exercise restraint and be selective. For instance, we do not treat Cahn's problem of wetting transitions, fascinating as the subject may be.[6] We rely on physical chemistry more than on statistical physics. In the same vein, we have elected not to cover the following topics:

- Superfluids, which are systems of exquisite elegance, but require of students the kind of technical maturity that only comes with experience
- Certain recent developments of a purely hydrodynamic nature, such as the inertial behavior of drops hitting a surface
- Wetting by volatile fluids
- The dynamic behavior of wetting in the presence of surfactants

Our task was not easy. Fortunately, we operated in marvelously stimulating environments, both at the Collège de France and at the Institut Curie, where several generations of experimenters and theorists had already done pioneering research on wetting. To mention but some of the founding members, we owe much to A. M. Cazabat, J. M. di Meglio, H. Hervet, F. Heslot, J. F. Joanny, L. Léger, T. Ondarçuhu, E. Raphaël, and F. Rondelez. We are also grateful to our outside friends, notably P. Pincus, Y. Pomeau, T. Witten, and M. Shanahan. They did not always embrace our views, but they did force us to think. We realize this book is far from perfect. But we did try to convey the sense of curiosity and joy that infused the members of our various research teams as they grappled for the past 20 years with drops large and small.

Acknowledgments: We owe a special debt of gratitude to our capable proofreaders, Christian Counillon, Annick Lesne, and Emilie Echalier, and to our "electronic editors," Florence Bonamy, Yvette Heffer, and Pierre-Henri Puech, assisted by Nicole Blandeau. Their willingness to invest so much of their time and energy in this project is enormously appreciated. Without the discipline they wisely imposed on us, this book would never have seen the light of day.

References

[1] H. Bouasse, *Capillarité et phénomènes superficiels* (Capillarity and Surface Phenomena) (Paris: Delagrave, 1924).

[2] B. Derjaguin, *Kolloid Zh* **17**, 191 (1955).

[3] K. Mysels, K. Shinoda, and S. Frankel, *Soap Films* (London: Pergamon, 1959).

[4] W. Zisman, Contact angles, wettability and adhesion, *Adv. in Chem. Series*, No. 43, Amer. Chem. Soc., Washington, D.C., ed. by F. M. Fowkes, p. 1.

[5] P. G. de Gennes, Wetting: Statics and dynamics, *Rev. Modern Physics* **57**, 827 (1985).

[6] D. Bonn and D. Ross, Wetting transitions, *Rep. Prog. Phys.* **64**, 1085 (2001).

1
Capillarity: Deformable Interfaces

Capillarity is the study of the interfaces between two immiscible liquids, or between a liquid and air. The interfaces are deformable: they are free to change their shape in order to minimize their surface energy. The field was created in the early part of the 19th century by Pierre Simon de Laplace (1749–1827) and Thomas Young (1773–1829). Henri Bouasse wrote a wonderful account of developments in capillarity in a book he published in 1924.[1] This discipline enables us to understand the games water can play to break the monotony of a rainy day or the tricks it performs while washing dishes. On a more serious note, capillarity plays a major role in numerous scientific endeavors (soil science, climate, plant biology, surface physics, and more), as well as in the chemical industry (product formulation in pharmacology and domestics, the glass industry, automobile manufacturing, textile production, etc.).

1.1 Surface Tension

A liquid flows readily; yet it can adopt extremely stable shapes. A drop of oil in water or a soap bubble forms a perfect sphere that is smooth on an atomic scale and is hardly deformable (Figure 1.1).[2] The fluctuations of the surface thickness are of the order of a mere Angström. A liquid surface can be thought of as a stretched membrane characterized by a surface tension that opposes its distortion.

We will focus our attention on the physical origin and consequences of the phenomenon of surface tension.

FIGURE 1.1. Drops and bubbles form perfect spheres.[2] (From *A Drop of Water: A Book of Science and Wonder*, by Walter Wick. Published by Scholastic Press, a division of Scholastic Inc. Photographs © 1997 by Walter Wick. Reproduced by permission.)

1.1.1 Physical Origin

A liquid is a condensed state in which molecules attract one another. When the attraction is stronger than thermal agitation, molecules switch from a gas phase to a phase that is dense, although still disordered—what we call a liquid. A molecule in the midst of a liquid benefits from interactions with all its neighbors and finds itself in a "happy" state. By contrast, a molecule that wanders to the surface loses half its cohesive interactions (Figure 1.2) and becomes "unhappy." That is the fundamental reason that liquids adjust their shape in order to expose the smallest possible surface area. When dry, your hair is likely to be full and thick, whereas the moment it gets wet, it sticks together in a drab, droopy mass. (Figure 1.3).

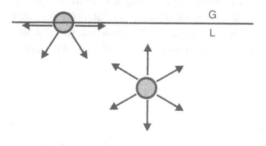

FIGURE 1.2. An "unhappy" molecule at the surface: It is missing half its attractive interactions.

FIGURE 1.3. Full dry hair vs. sticky wet hair.

TABLE 1.1. Surface tension of a few common liquids (at 20°C unless otherwise noted) and interfacial tension of the water/oil system.

Liquid	Helium (4K)	Ethanol	Acetone	Cyclohexane	Glycerol
γ(mN/m)	0.1	23	24	25	63
Liquid	Water	Water (100°C)	Molten glass	Mercury	Water/oil
γ(mN/m)	73	58	~300	485	~ 50

When segregated to the surface, a liquid molecule is in an unfavorable energy state. If the cohesion energy per molecule is U inside the liquid, a molecule sitting at the surface finds itself short of roughly $U/2$. The surface tension is a direct measure of this energy shortfall per unit surface area. If a is the molecule's size and a^2 is its exposed area, the surface tension is of order $\gamma \cong U/(2a^2)$. For most oils, for which the interactions are of the van der Waals type, we have $U \cong kT$, which is the thermal energy. At a temperature of 25°C, kT is equal to 1/40 eV, which gives $\gamma = 20$ mJ/m^2.

Because water involves hydrogen bonds, its surface tension is larger ($\gamma \approx 72$ mJ/m^2). For mercury, which is a strongly cohesive liquid metal, $U \approx 1$ eV and $\gamma \approx 500$ mJ/m^2. Note that γ can equivalently be expressed in units of mN/m.

Likewise, the surface energy between two non-miscible liquids A and B is characterized by an interfacial tension γ_{AB}. Table 1.1 lists the surface tensions of some ordinary liquids (including those used in the experiments to be described in the course of these chapters), as well as the interface tension between water and oil.

Although its origin can be explained at the molecular level, the surface tension γ is a macroscopic parameter defined on a macroscopic scale, as we will see shortly.

1.1.2 Mechanical Definition: Surface Energy and Capillary Force

Surface Work

It is well known that supplying energy is necessary to create surfaces. That fact is plainly obvious when you beat egg whites into a meringue or when you make an emulsion of water in oil while preparing a mayonnaise.

Suppose one wants to distort a liquid to increase its surface area by an amount dA. The work required is proportional to the number of molecules that must be brought up to the surface, i.e., to dA; and one can write:

$$\delta W = \gamma \cdot dA \tag{1.1}$$

where γ is the surface (or interfacial) tension. Dimensionally, $[\gamma] = EL^{-2}$. The surface tension γ is thus expressed in units of mJ/m^2. Stated in words,

> γ is the energy that must be supplied to increase the surface area by one unit.

Surface tension also contributes to thermodynamic work.[3] It can be defined as the increase in internal energy U or in free energy F that accompanies an increase in surface area:

$$\gamma = \left[\frac{\partial F}{\partial A}\right]_{T,V,n} \tag{1.2}$$

where n is the number of molecules and V is the total volume.

Surface thermodynamics is a rather subtle science, which we will refrain from reviewing here. The interested reader is urged to consult the text by Rowlinson and Widom.[3] We simply note that, if one works with a fixed chemical potential μ, it is convenient to use the grand potential $\Omega = F - n\mu = -pV + \gamma A$.

Capillary Forces

Surface tension can also be viewed as a force per unit length. Dimensionally, one can write $[\gamma] = FL^{-1}$, and one can express γ in units of N/m. We proceed to describe a few experiments in which γ manifests itself as a force (Figure 1.4).

1. Imagine a rigid metal frame bent in the form of a wedge and its two extremities connected by a thin sewing thread. If one deposits a liquid film (such as a soap film) within the wedge, the film will want to shrink its surface area. As it does so, it pulls perpendicularly and uniformly on every element of the thread, which will ultimately lose its slack and take on the shape of a taut circular arc.
2. Consider a rigid frame supporting a liquid membrane. A flexible loop, secured by two threads to the frame, is embedded in the membrane (Figure 1.4). The loop is free to take on any shape until the membrane is punctured, at which time the loop stretches into a circle.
3. Visualize a glass rod bent to form three sides of a rectangle. A second rod, free to roll on the two parallel sides of the rectangle, constitutes the fourth side of length l (Figure 1.4). The apparatus is dipped into a glyceric liquid (containing water, bubble soap, and glycerine to make the mixture viscous). As soon as the apparatus is removed from the liquid, one observes that the mobile rod moves spontaneously in the direction of the arrow so as to decrease the surface area of the liquid. If the frame is tilted, it is even possible for the mobile rod to climb up the incline, only to fall back down suddenly the moment the liquid membrane is pierced.

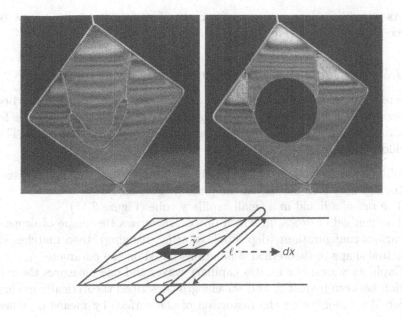

FIGURE 1.4. Manifestation of surface tension: force normal to the line (wire, rod). (From *A Drop of Water: A Book of Science and Wonder*, by Walter Wick. Published by Scholastic Press, a division of Scholastic Inc. Photographs © 1997 by Walter Wick. Reproduced by permission.)

If the mobile rod moves by a distance dx, the work done is

$$\delta W = F \cdot dx = 2\gamma \cdot l \cdot dx \qquad (1.3)$$

where the factor of 2 reflects the presence of two interfaces. This demonstrates that γ is also the force exerted per unit length of the rod. In conclusion,

$\vec{\gamma}$ is a force (per unit length) normal to the rod in the plane of the surface and directed toward the liquid.

Capillary forces are truly remarkable. They enable insects to walk on water. However, should the pond become polluted with detergents, which lower the surface tension, the unfortunate insects will drown! This phenomenon of flotation can be studied by depositing a sewing needle on a very thin piece of toilet paper brought up the surface. After gently removing the toilet paper, the needle keeps on floating. The moment one adds a drop of detergent, the needle sinks instantly.

As we will see later, these two aspects of surface tension—energy and force—will be a recurring theme.

1.1.3 Measurements of Surface (or Interfacial) Tensions

There exist numerous measurement techniques, all of which have been described in a book by A. W. Adamson.[4] It is worth mentioning a few standard methods, which will be discussed in detail in chapter 2. They include

- Wilhelmy's method, in which one dips a thin plate or a ring and measures the capillary force acting on the plate (Figure 2.24),
- The rise of a liquid in a small capillary tube (Figure 2.17),
- The method of drops, in which one characterizes the shape of drops in various configurations (deposited, rotating, hanging), then matches the actual shape to theoretical simulations based on the parameter γ,
- Capillary waves: one excites capillary waves and one measures the relation between frequency and wavelength (described theoretically in chapter 5) by monitoring the distortion of the surface by means of a laser beam.

Every one of these methods requires considerable precautions. Liquid surfaces are ideal, smooth on an atomic scale, and chemically homogeneous. Unfortunately, they are easily contaminated. Early data on the surface tension of water were plagued with enormous variations until the day (about one hundred years ago) when Agnès Pockels, while experimenting in her kitchen, realized that it was necessary to "scrape" the surface of water. A surface of fresh water has a well-defined surface tension $\gamma = 72$ mN/m. But water, whose surface tension is particularly high, happens to easily get contaminated, which lowers its surface tension. To avoid this serious drawback and prepare liquid surfaces that do not change with time, one often uses silicone-based oils, which have a low surface tension ($\gamma \approx 20$ mN/m). For the same reason, these substances are also used as anti-graffiti and anti-stain agents to protect the facades of buildings: they make the surface non-adhesive.

1.1.4 Laplace Pressure

This section draws on the work of Laplace published in 1805.[5] Surface tension is at the origin of the overpressure existing in the interior of drops and bubbles. This pressure difference has multiple consequences. For instance, smaller drops will disappear in favor of larger ones in an emulsion, and they will be the first to evaporate during the cooling phase of an aerosol. The pressure difference also explains the phenomenon of capillary adhesion between two plates, between hairs or fibers, or in wet sand, all of which are induced by capillary bridges.

FIGURE 1.5. Overpressure inside a drop
of oil "o" in water "w."

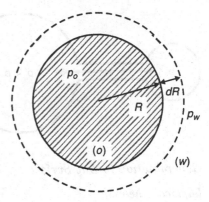

As one passes across a curved surface or interface, a jump in pressure
occurs, which we proceed to evaluate, first for a sphere, and then for any
curved surface.

Sphere

We take the example of a drop of oil (o) in water (w) (Figure 1.5). In order
to lower its surface energy, the drop adopts a spherical shape of radius R.
If the o/w interface is displaced by an amount dR, the work done by the
pressure and capillary force can be written as

$$\delta W = -p_o\, dV_o - p_w\, dV_w + \gamma_{ow}\, dA \tag{1.4}$$

where $dV_o = 4\pi R^2\, dR = -dV_w$, and $dA = 8\pi R\, dR$ are the increase in volume and surface, respectively, of the drop, p_o and p_w are the pressures in
the oil and water, and γ_{ow} is the interfacial tension between oil and water.
The condition for mechanical equilibrium is $\delta W = 0$, which amounts to

$$\Delta p = p_o - p_w = \frac{2\gamma_{ow}}{R}. \tag{1.5}$$

For an aerosol drop of radius 1 μm, Δp is typically comparable to the
atmospheric pressure. Note that equation (1.5) can be obtained just as
well by minimizing the grand potential $\Omega = -p_o V_o - p_w V_w + \gamma_{ow} A$.

The smaller the drop, therefore, the greater its inner pressure. This property can be verified with soap bubbles. By connecting two bubbles (Figure 1.6) of different sizes, one can readily observe that the smaller one
empties itself into the larger one. In an emulsion of oil in water, small
drops disappear in favor of large ones because of this overpressure, which
makes them thermodynamically unstable (a phenomenon known as Ost-
wald's ripening).

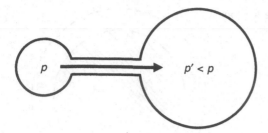

FIGURE 1.6. Small bubbles empty themselves into larger ones.

Generalization to Any Surface

Laplace's theorem:

The increase in hydrostatic pressure Δp that occurs upon traversing the boundary between two fluids is equal to the product of the surface tension γ and the curvature of the surface $C = \frac{1}{R} + \frac{1}{R'}$:

$$\Delta p = \gamma \left(\frac{1}{R} + \frac{1}{R'} \right) = \gamma C \qquad (1.6)$$

where R and R' are the radii of curvature of the surface.

As was the case for the sphere, equation (1.6) can be demonstrated by calculating the work done by the forces of pressure and the capillary forces during an infinitesimally small displacement or, alternately, by minimizing the grand potential.[4]

A convenient way to illustrate how to measure the curvature of a surface is to use the example of a pear (Figure 1.7). The curvature at point M is determined by inserting a needle defining the direction $\vec{N}$ normal to the surface. Next, the pear is cut along two mutually orthogonal planes

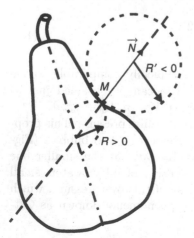

FIGURE 1.7. Measuring the curvature of a pear at a particular point.

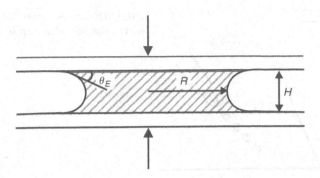

FIGURE 1.8. Capillary adhesion of two plates with a drop squeezed in-between.

intersecting each other along $\vec{N}$. The intersection of these planes with the surface of the pear defines two curves, the radii of curvature of which are R and R'. Note that R and R' are to be treated as *algebraic* quantities: R is defined as positive if the center of the corresponding circle lies *inside* the pear, and negative otherwise. A remarkable property of the curvature C is that it is independent of the orientation of the planes. If there exists a symmetry axis and one of the two planes contains that axis, the corresponding R and R' are then referred to as the *principal* radii of curvature.

Capillary Adhesion

Two wetted surfaces can stick together with great strength if the liquid wets them with an angle $\theta_E < \pi/2$. The angle θ_E is defined in Figure 1.8. (It will be discussed in more detail in Section 1.2.) Imagine that we mash a large drop between two plates separated by a distance H. The drop forms what is called a *capillary bridge* characterized by a radius R and a surface area $A = \pi R^2$. The Laplace pressure within the drop reads

$$\Delta p = \gamma \cdot \left(\frac{1}{R} - \frac{\cos \theta_E}{H/2} \right) \approx -\frac{2\gamma \cos \theta_E}{H}. \qquad (1.7)$$

The force that glues the two plates together is attractive as long as $\theta_E < \pi/2$. If $H \ll R$, it is equal to

$$F = \pi R^2 \frac{2\gamma \cos \theta_E}{H}.$$

For water, using $R = 1$ cm, $H = 5$ μm, and $\theta_E = 0$ (best case), one calculates a pressure drop $\Delta p \sim 1/3$ atm and an adhesive force $F \sim 10$ N, which is enough to support the weight of one liter of water!

1.1.5 Minimal Surfaces

We have just seen that a liquid evolves spontaneously so as to minimize its surface area, and we illustrate this property in Figure 1.9. At equilibrium, minimum area surfaces satisfy Laplace's equation.

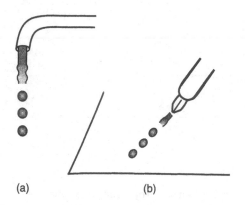

FIGURE 1.9. A stream of water and a trace of ink break up into droplets.

(a)   (b)

1.1.5.1 Jet

Upon opening a faucet, it is common to observe the water stream breaking up into droplets (Figure 1.9a) in order to lower its surface energy. To understand better the benefits of a lower surface area, we may mentally break up a cylinder of radius R and length L into n droplets of radius r. Conservation of volume dictates that

$$\pi R^2 L = \frac{4}{3}\pi r^3 n. \tag{1.8}$$

Let us examine the ratio of the final surface area S_n of the drops to the initial surface area S_0 of the cylinder. After eliminating n with the help of equation (1.8), we obtain

$$\frac{S_n}{S_0} = \frac{n \times 4\pi R^2}{2\pi RL} = \frac{3R}{2r}. \tag{1.9}$$

It is clear that the surface area of the drops is less than that of the original cylinder as soon as $r > \frac{3}{2}R$.

Plateau was the first to understand that the cylinder distorts itself spontaneously in order to lower its surface energy as soon as the wavelength λ_d of the distortion exceeds the perimeter of the cylinder.[6] The distortion then amplifies itself and the liquid cylinder fragments into drops. Some time later, Lord Rayleigh showed that the size of the drops was determined by the fastest distortion mode ($\lambda_d/2R \approx 4.5$ in the inertial regime).[7] This instability came to be named after Rayleigh. A liquid stream is one way to produce emulsions of uniform size. The technique is used in the manufacture of homogenized milk, as well as in many other industrial processes. We will return to the Plateau-Rayleigh instability in more detail when we describe the instability of liquid sheaths on fibers in chapter 5.

If one draws a line of ink on a piece of plastic, the line breaks up into droplets because, for exactly the same reason, a section of cylinder is less stable than a string of spherical caps. In a subsequent chapter devoted to the dynamics of wetting, we will see how this phenomenon controls numerous hydrodynamic instabilities that show up when a liquid flows and collects in a ridge in the vicinity of the line marking the boundary between liquid, air, and substrate. A very simple experiment consists in depositing oil on half a Teflon pan and tilting it. The boundary line becomes wavy as fingers (or run-offs) begin to grow and expand.

1.1.5.2 Drop on a Fiber

Consider a drop of radius R deposited on a fiber of radius b (b is typically 10 to 100 μm). Assume that the liquid is able to wet the fiber, which means that the two media will connect smoothly at an angle equal to zero. The fiber may be a strand of hair, a textile thread, or a thin glass fiber. The shape taken on by the drop is sketched in Figure 1.10. Since its radius R is very much greater than b, the overpressure within the drop remains low ($\Delta p \approx 2\gamma/R$). At the outer point of contact between the drop and the fiber, one of the radii of curvature becomes very small (equal to b). Therefore, the other radius of curvature must become negative (of the order of $-b$) in order for the total radius of curvature to remain small.

We may calculate the details of the drop's profile $z(x)$. The surface has a constant curvature throughout given by

$$\frac{1}{R_1} + \frac{1}{R_2} = C = \frac{\Delta p}{\gamma} \tag{1.10}$$

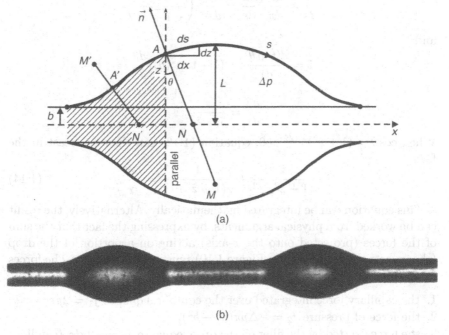

(a)

(b)

FIGURE 1.10. Drop deposited on a fiber. Schematic cross section (a); photograph of silicone oil drops deposited on a carbon fiber (b).

where Δp is the overpressure within the drop. The radii R_1 and R_2 at point A are $R_1 = AM$ (M is the center of curvature of the meridian curve in the plane of the figure) and $R_2 = AN$ (in the perpendicular plane). Point N lies at the intersection of the normal to the surface at point A and the symmetry axis. R_2 is always positive, while R_1 is positive if points M and N are on the same side of A, and negative if M switches to the other side. As an example, at point $A', R_1' = A'M'$ is negative, and $R_2' = A'N'$ is positive.

If s is the curvilinear coordinate along the meridian curve oriented from left to right and θ is the angle between the normal to the drop's surface and the vertical direction, R_1 and R_2 are given by

$$z = R_2 \cos \theta \tag{1.11}$$

$$ds = -R_1 d\theta. \tag{1.12}$$

Equation (1.10) leads to

$$-\frac{d\theta}{ds} + \frac{\cos \theta}{z} = \frac{\Delta p}{\gamma}. \tag{1.13}$$

Along the meridian curve $z(x)$, we have $dz = ds \sin \theta$ and $dx = ds \cos \theta$. With the notation $\dot{z} = \frac{dz}{dx}$ and $\ddot{z} = \frac{d^2 z}{dx^2}$, we get $ds = dx \sqrt{1 + \dot{z}^2}$ (ds and dx always have the same sign), from which it follows that

$$\frac{d\theta}{ds} = \frac{d\theta}{dx} \cdot \frac{dx}{ds} = \frac{d\theta}{dx} \cdot \left(\frac{1}{\sqrt{1 + \dot{z}^2}} \right)$$

and

$$\ddot{z} = \frac{d(\tan \theta)}{dx} = (1 + \tan^2(\theta)) \frac{d\theta}{dx}$$

or

$$\frac{d\theta}{dx} = \frac{\ddot{z}}{1 + \dot{z}^2}.$$

When $\cos \theta = (1 + \dot{z}^2)^{-1/2} > 0$, equation (1.13) can then be recast in the form

$$-\frac{\ddot{z}}{(1 + \dot{z}^2)^{3/2}} + \frac{1}{z(1 + \dot{z}^2)^{1/2}} = \frac{\Delta p}{\gamma}. \tag{1.14}$$

This equation can be integrated mathematically. Alternatively, the result can be worked from physical arguments, by expressing the fact that the sum of the forces (projected onto the x-axis) acting on a portion of the drop (shown as a shaded section in Figure 1.10) must add up to zero. The forces in question are

1. the capillary force integrated over the contour, equal to $f_1 = 2\pi z \gamma \cos \theta$,
2. the force of pressure $f_2 = -\Delta p \pi (z^2 - b^2)$,
3. the force exerted by the fiber on the drop, equal to $f_3 = -2\pi b \gamma$ (capillary force on the inner contour).

The condition $f_1 + f_2 + f_3 = 0$ leads to

$$\frac{z}{\sqrt{1 + \dot{z}^2}} - \frac{\Delta p}{2\gamma}(z^2 - b^2) = b. \qquad (1.15)$$

The maximum radius of the drop is obtained when $\dot{z} = 0$ and $z = L$. Equation (1.15) then gives

$$\frac{\Delta p}{2\gamma}(L^2 - b^2) = L - b. \qquad (1.16)$$

For a large drop, the overpressure $\Delta p = \frac{2\gamma}{L+b}$ is roughly equal to the Laplace pressure for a drop of radius L since the correction term b then is negligible.

1.1.6 Minimal Surfaces With Zero Curvature

Henri Poincaré made some key contributions in this area.[8]

Meniscus on a Fiber

Consider an experiment in which one dips a fiber into a liquid bath. Our goal is to study the rise of the liquid while neglecting the influence of gravity (Figure 1.11). The liquid is assumed to wet the fiber.

There is a difference between the present situation and the previous experiment where a drop was deposited on a fiber. Here the liquid in the meniscus is in equilibrium with the liquid bath. As a result, we have $\Delta p = 0$, which defines a surface with zero curvature. The profile is given by equation (1.15), which reduces to

$$\frac{z}{\sqrt{1 + \dot{z}^2}} = b. \qquad (1.17)$$

This last equation could be obtained directly by noting that the vertical projection of the tension forces must be conserved. At a height x, the tension force is $2\pi z\gamma \cos\theta = 2\pi\gamma b$. Since $\tan\theta = \dot{z}$, equation (1.17) is readily

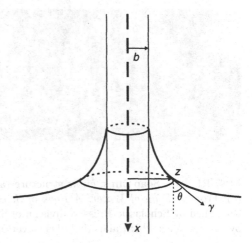

FIGURE 1.11. Water climbing up a glass fiber of radius $b \approx 10$ μm.

recovered. The profile of the drop is that of a hanging chain, known as a *catenary curve:*

$$z = b \cdot \cosh\left(\frac{x}{b}\right). \tag{1.18}$$

Soap Film

Let us reenact Plateau's experiment, which consists in stretching a soap film (glyceric liquid) between two circular rings of radius R. The rings can be fashioned out of heavy-duty copper wire hammered flat. One subsequently dips them into the soap mixture and then gently pulls them apart (Figure 1.12b). The distance between the two rings is $2D$. Since the pressure is the same inside and outside, the surface of the film has zero curvature ($R_1^{-1} + R_2^{-1} = 0$). As the rings are pulled ever farther apart, the surface area of the film between the rings stretches until the film bursts when $R/D \approx 1.5$. Plateau was the first to study the surface created between the two rings. The profile $r(x)$ is that of a surface of revolution with zero curvature, which satisfies equation (1.13) where Δp is set equal to 0. In accordance with equation (1.18), the profile of the liquid must adopt the shape of a catenary curve that connects smoothly with the rings, at which point $r(x = \pm D) = R$. If R_m is the radius of the circle at the waist (where $x = 0$), then the profile is

$$r(x) = R_m \cdot \cosh\left(\frac{x}{R_m}\right). \tag{1.19}$$

When $x = D$, we have

$$\frac{R}{R_m} = \cosh\left(\frac{D}{R_m}\right). \tag{1.20}$$

This equation has two solutions for R_m, corresponding to two surfaces, both with zero curvature. The surface area is minimum for the first and maximum for the second.

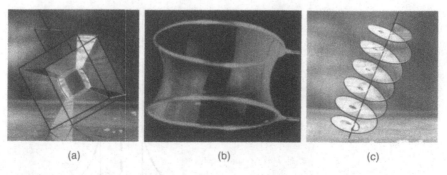

(a)   (b)   (c)

FIGURE 1.12. Soap films: cubic structure (a); catenary (b); spiral (c); (a) and (c): From *A Drop of Water: A Book of Science and Wonder,* by Walter Wick. Published by Scholastic Press, a division of Scholastic Inc. Photographs © 1997 by Walter Wick. (b): Palais de la Découverte. Reproduced by permission.

For a critical value of the ratio R/D ($R/D = 1.509$), the two solutions become identical. For $R/D \leq 1.509$, solutions no longer exist and the film bursts.

With more complex frames, one can generate a variety of minimal surfaces with zero curvature. Figure 1.12 shows a cubic structure, a hanging chain, and a spiral.

1.2   Contact Between Three Phases: Wetting

Wetting refers to the study of how a liquid deposited on a solid (or liquid) substrate spreads out. The phenomenon is pertinent to numerous industrial areas, a few of which are listed below:

- chemical industry (paints, ink, coloring ingredients, insecticides),
- automobile manufacturing (surface preparation prior to painting, treatment of glass to prevent water from dewetting, treatment of tires to promote adhesion even on wet or icy roadways),
- glass (anti-stain or anti-frost treatment),
- food (dissolving powders such as milk or cocoa),
- soil science (penetration of liquids into porous rocks),
- construction (waterproofing of concrete, protection of monuments, treatment of greenhouse plastic),
- domestics (spreading of creams, application of mascara to eyelashes, self-drying shampoos).

It also plays a role in the life sciences. A few notable examples follow:

- inflation of lungs at birth initiated by surfactant molecules that lower the surface energy of the lungs. In some premature babies, these molecules are missing and the lungs are not ready to function on their own. This respiratory stress syndrome, known as hyaline membrane disease, is alleviated by the swift delivery of suitable surfactants),
- rise of sap in plants,
- locomotion of insects on the surface of water,
- adhesion of parasites on wet surface (e.g., pyriculariosis of rice, or rice blast),
- wetting of the eye. The cornea is by nature very hydrophobic, yet a normal eye is wet! Proteins (called mucins), present in tears, turn the surface of the eye hydrophilic, stabilizing the lachrymal film. If one accidentally smears a fatty cream on the eye, it dries up, causing considerable discomfort. Some individuals happen to suffer from "dry eyes" and must apply artificial tears to compensate for their natural deficiency in mucins.

"Understanding" wetting enables us to explain why water spreads readily on clean glass but not on a plastic sheet. "Controlling" it means being able to modify a surface to turn a non-wettable solid into one that is wettable (plastic covered with a layer of gold, cornea coated with mucin), or, vice versa. For instance, it is possible to turn glass just as non-wetting as Teflon by depositing a thin coating of fluorinated molecules on it.

In the following section, we begin by characterizing two types of wetting:

- total wetting, when the liquid has a strong affinity for the solid; and
- partial wetting, in the opposite case.

Next we will describe criteria useful for predicting whether or not a liquid will wet a particular substrate. We will show how a simple monolayer deposited on the substrate can reverse the behavior of the interface, i.e., change it from wetting to non-wetting and vice versa. We will pay particular attention to those liquids and solids that have been used to implement well-controlled systems and we will describe the most common surface treatments used in the physical chemistry of wetting. The modifications involved will be discussed in chapter 2.

1.2.1   Two Types of Wetting: The Spreading Parameter S

When a water drop is placed down on very clean glass, it spreads completely. By contrast, the same drop deposited on a sheet of plastic remains stuck in place. The conclusion is that there exist two regimes of wetting depicted in Figure 1.13. The parameter that distinguishes them is the so-called *spreading parameter S*, which measures the difference between the surface energy (per unit area) of the substrate when dry and wet:

$$S = [E_{substrate}]_{dry} - [E_{substrate}]_{wet} \qquad (1.21)$$

or

$$S = \gamma_{SO} - (\gamma_{SL} + \gamma), \qquad (1.22)$$

where the three coefficients γ are the surface tensions at the solid/air, solid/liquid, and liquid/air interfaces, respectively.

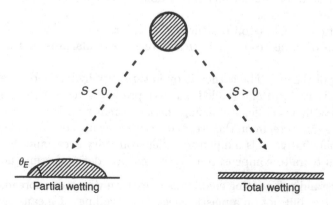

FIGURE 1.13. The two wetting regimes for sessile drops.

S > 0: Total Wetting

If the parameter S is positive, the liquid spreads completely in order to lower its surface energy ($\theta_E = 0$). The final outcome is a film of nanoscopic thickness resulting from competition between molecular and capillary force (see chapter 4).

S < 0: Partial Wetting

The drop does not spread but, instead, forms at equilibrium a spherical cap resting on the substrate with a contact angle θ_E. A liquid is said to be "mostly wetting" when $\theta_E \leq \pi/2$, and "mostly non-wetting" when $\theta_E > \pi/2$. Note, however, that $\theta_E = \pi/2$ plays no particularly significant role from a thermodynamical standpoint, in contrast to $\theta_E = 0$, which corresponds to a condition of wetting transition. We will see in chapter 2 that a "mostly wetting" liquid spontaneously invades a capillary, a porous medium, or a sponge.

Law of Young-Dupré

The contact angle can be obtained via one of two methods (Figure 1.14):

1. The first method consists in tallying up the capillary forces acting on the line of contact (also called triple line) and equating the sum to zero. When normalized to a unit length, these forces are the interface tensions between the three phases (S/L/G). By projecting the equilibrium forces onto the solid plane, one obtains Young's relation (which he derived in 1805):[9]

$$\gamma \cos \theta_E = \gamma_{SO} - \gamma_{SL} \qquad (1.23)$$

Substituting (1.22) into (1.23) yields:

$$S = \gamma(\cos \theta_E - 1)$$

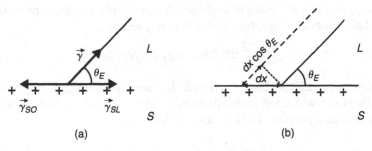

(a)   (b)

FIGURE 1.14. Determination of θ_E: (a) via forces or (b) via works.

It is evident that θ_E can be defined only if the spreading parameter is negative. θ_E increases when the liquid is non-wetting.

The projection of the capillary forces onto the vertical axis is balanced out by a force of reaction exerted by the solid. If the solid is hard, no distortion is observable. If, on the other hand, it is soft (e.g., rubber or a coat of paint), it does distort. That is the reason why a water drop left on a fresh coat of paint leaves behind a circular mark.

2. The second method relies on calculating the work done by moving the line of contact by a distance dx:

$$\delta W = (\gamma_{SO} - \gamma_{SL})\, dx - \gamma \cos \theta_E \, dx \qquad (1.24)$$

This work is equal to zero at equilibrium, which indeed leads to equation (1.23).

Measuring Contact Angles

There are several methods for measuring θ_E, which will be discussed in chapter 2. For relatively large angles, it is possible to take a side-view photograph of the profile and use the snapshot to determine the angle. For better precision and for angles less than $\pi/4$, an optical reflection technique is often used. The drop is illuminated by a collimated laser beam which, upon reflection, becomes divergent. The divergence angle is related to the contact angle. For small angles and a higher precision still, an interference method is preferred; it relies on monitoring the constant-thickness fringes generated by a liquid wedge. Finally, a less accurate method, but one that is useful when studying the dynamics of wetting (when the dynamical contact angle θ_D is different from the static angle θ_E), consists in recording the distortion of the image of a grid seen through the liquid wedge.

1.2.2   Wetting Criteria: Zisman's Rule

Is it possible to predict whether a solid surface is wettable? Fortunately, the answer is yes.[10, 11] Surfaces belong in one of two categories:

1. "High-energy" (HE) surfaces are those for which the chemical binding energy is of the order of 1 eV, on which nearly any liquid spreads. High-energy surfaces are made of materials that are ionic, covalent, or metallic. In this category, the interface tension is given by

$$\gamma_{SO} \approx \frac{E_{binding}}{a^2} \sim 500 - 5{,}000 \text{ mN/m}. \qquad (1.25)$$

2. Low-energy (LE) surfaces, for which the chemical binding energy is of the order of kT, which are generally hardly wettable. They include molecular crystals and plastics. In this case, we have

$$\gamma_{SO} \approx \frac{kT}{a^2} \sim 10 - 50 \text{ mN/m}. \qquad (1.26)$$

Actually, the surface energy γ_{SO} in contact with air is not altogether sufficient to predict wettability. What is in fact needed is the sign of the spreading parameter S given by

$$S = \gamma_{SO} - (\gamma_{SL} + \gamma).\qquad(1.27)$$

We shall now restrict our attention to an idealized (but important) case—the interactions (liquid–liquid and liquid–solid) are purely of the van der Waals type. It is then possible to relate S to the **electric polarisabilites** (α_S, α_L) of S and L.[12,13]

The approach is schematically illustrated in Figure 1.15. To estimate γ_{SO}, one brings together two semi-infinite solid media. At first, the energy is $2\gamma_{SO}$. Upon merging them, one gains the van der Waals energy V_{SS} (per unit area). The latter energy is related to the polarizability α_S of the solid via the relation $V_{SS} = k \cdot \alpha_S^2$, where k is a constant. The surface energy of the resultant solid is zero. Hence,

$$2\gamma_{SO} - V_{SS} = 0.\qquad(1.28)$$

To estimate γ_{SL}, one brings the solid and the liquid together. One starts with an energy $\gamma + \gamma_{SO}$, and one picks up the van der Waals interactions V_{SL} between the solid and the liquid (as before, we have $V_{SL} = k\alpha_S\alpha_L$). This leads to

$$\gamma_{SL} = \gamma + \gamma_{SO} - V_{SL},\qquad(1.29)$$

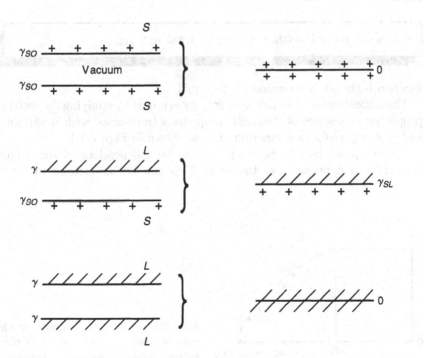

FIGURE 1.15. Determining the interface energies γ_{ij} by bonding i and j together.

To estimate γ, one brings two semi-infinite liquid media together. One starts with 2γ and then gains the van der Waals interaction V_{LL}, which yields

$$2\gamma - V_{LL} = 0. \qquad (1.30)$$

Equations (1.28)–(1.30) can now be combined to produce an estimate of the spreading parameter S:

$$S = \gamma_{SO} - (\gamma_{SL} + \gamma) = V_{SL} - V_{LL} = k(\alpha_S - \alpha_L)\alpha_L. \qquad (1.31)$$

It becomes clear that the wettability criterion is not γ_{SO}, since that quantity drops out. What does matter is the sign of S. If $\alpha_S > \alpha_L$, S is positive and wettability is total. Hence the rule

> *A liquid spreads completely if it is less polarizable than the solid.*

This explains why liquid helium, with its extremely low polarizability, spreads on most solids.

All liquids spread on glass, metals, and ionic crystals. By contrast, wetting may be total or partial on plastics and molecular crystals, depending on which specific liquid is used. The empirical criterion worked out by Zisman allows us to classify solids. Each solid substrate has a critical surface tension γ_C such that

> $\gamma > \gamma_C \Rightarrow$ partial wetting; $\gamma < \gamma_C \Rightarrow$ total wetting

where γ is the surface tension of the liquid.

The critical surface tension γ_C can be determined by studying the wetting properties of a series of chemical compounds (n-alkanes, with n variable) and plotting $\cos\theta_E$ as a function of γ, as shown in Figure 1.16.

For non-polar liquids, γ_C turns out to be independent of the liquid! Rather, it is a property of the solid. This can be understood in terms

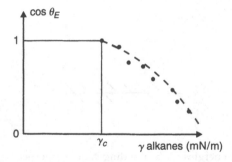

FIGURE 1.16. Determination of the critical tension γ_C of a sheet of plastic by means of a series of alkanes.

TABLE 1.2. Critical surface tension of a few solid polymers.

Solid	Nylon	PVC	PE	PVF$_2$	PVF$_4$
γ_C (mN/m)	46	39	31	28	18

of equation (1.31), which shows that $S(\gamma_C) = 0$ (by definition of γ_C) is satisfied when $\alpha_L = \alpha_S$, α_S being a characteristic of the solid. Indeed, we have $S(\gamma_C) = k(\alpha_S - \alpha_L)\alpha_S = 0$ when $\alpha_S = \alpha_L$. When $\gamma < \gamma_C$, we have $S > 0$ and wetting is total. When $\gamma > \gamma_C$, we have $S < 0$ and wetting is partial.

The method just described makes it possible to characterize not only the surface of non-wettable solids (see Table 1.2), but also the surfaces of HE solids made non-wettable by a suitable surface treatment. For example, glass covered with a fluorinated molecular coating can have a critical surface tension γ_C as low as 10 mN/m, rather than the 150 mN/m characteristic of clean glass

1.2.3 Choice of Solid/Liquid Pairs

We now discuss a few of the characteristics of selected liquids and solids that are often used in wetting experiments. The primary emphasis is on those materials that are suitable for well-controlled experiments ("ideal" liquids and solids).

1.2.3.1 Ideal Liquids

Careful experiments often rely on the following categories of fluids:

- *Pure and non-volatile liquids*, to avoid the Marangoni effects related to evaporation (see chapter 10): Hydrogenated and fluorinated silicone-based oils, long alkanes.
- *Liquids of the "van der Waals" type*, in which the analysis of long-range forces simplifies itself (see chapter 4),
- *Liquids with low surface tension*, which are relatively immune to self contamination;
- *Liquids with an adjustable viscosity coefficient* η (*usually polymer melts with different chain lengths*). They are useful for studying time-dependent phenomena. A characteristic velocity $V^* = \gamma/\eta$ controls the dynamics of wetting. That velocity can range from about 1 μm/s to 70 m/s (for water).

PDMS

Polydimethylsiloxanes (or PDMSs) are silicone oils that readily comply with the criteria listed above (Table 1.3). They are routinely used in numerous industrial applications such as lubricating agents and waterproofing

TABLE 1.3. Main characteristics of selected PDMSs at ambient temperature.
η is the viscosity, ρ the density, κ^{-1} the capillary length, and $V^* = \gamma/\eta$ the
characteristic liquid velocity.

Molecular Mass (g)	η (mPa-s)	ρ (kg/m^3)	γ (mN/m)	κ^{-1} (mm)	V^* (mm/s)
3780	48	960	20.8	1.49	433
9430	193	968	21.0	1.49	109
28,000	971	971	21.2	1.49	22
62,700	11,780	974	21.5	1.5	1.8
204,000	293,100	977	21.5	1.5	0.07

compounds (paper, textiles, and anti-foam agents). Their general formula
is $(CH_3)_3-Si-O-[(CH_3)_2SiO]_n-Si(CH_3)_3$.

PDMSs consist of a siloxane skeleton (Si–O group) linked to two methyl
groups. These groups are responsible for the non-polar and hydrophobic
characters of PDMSs and their great thermal stability, as well as their
optical transparency. The number n of monomer units is called the degree
of polymerization.

PDMSs have a number of noteworthy characteristics:

- The chains are highly flexible and the corresponding oils are fluid at
  ambient temperature. The glass transition temperature T_g is $-128°C$.[14]
- Even for relatively small values of the number n of monomers, the vapor
  pressure is quite low. These substances are therefore non-volatile liquids.
- Their surface tension γ is low and practically independent of the molecular weight (see Table 1.3). γ decreases with temperature, typically at a
  rate 0.1 mN/m per degree.
- The viscosity η depends very strongly on the molecular weight, increasing by a factor of 6,000 as the molecular weight goes from 3,780 g to
  204,000 g. It decreases slowly with increasing temperature. The linear
  temperature coefficient is of the order of 10^{-2} K^{-1}, giving a dependence
  of the type $\eta(T) = \eta(T_0) \cdot [1 - 10^{-2}(T - T_0)]$.

Alkanes

Alkanes are made of a carbon chain terminated by a methyl group at either
end. The basic formula of a linear alkane containing n carbon atoms is
C_nH_{2n+2}.

The alkanes listed in Table 1.4 range from nonane ($n = 9$) to hexadecane
($n = 16$). They are liquids at 25°C and non-volatile (the saturation vapor
pressure of nonane at 20°C is 5 mbar). These liquids are stable and non-
polar. Therefore, they do not react with the surface of a substrate.

Alkanes have a low viscosity, of the order of 1 mPa-s (comparable to
that of water). The viscosity increases with the length of the chain, as
does the surface tension. Consequently, the contact angle between a drop

TABLE 1.4. Primary characteristics of a few alkanes (same notation as in Table 1.3).

Number n of carbon atoms	η (mPa-s)	γ (mN/m)	κ^{-1} (mm)	V^* (m/s)
9	0.71	22.9	1.8	32
10	0.92	23.9	1.8	26
12	1.35	25.4	1.9	19
16	3.34	27.6	1.9	8

of alkane and a given solid is an increasing function of the number n of carbon atoms (in a partial wetting regime) as observed in Figure 1.16. The characteristic velocity $V^* = \gamma/\eta$ is of the order of 1 m/s. Therefore, the observable dynamical processes are quite rapid.

1.2.3.2   Solid Substrates

Smooth Substrates or Substrates with a Controlled Roughness

To avoid hysteresis effects, it is advisable to use surfaces that are smooth on an atomic scale. Excellent candidates are

- silicon wafers of the type used in the microelectronics industry,
- floated glass, produced by flowing molten glass on liquid tin, which generates surfaces with a liquid-like smoothness (the technique is known as the Pilkington process),
- elastomers obtained by cross-linking a liquid film or a drop.

At the other extreme are fractal surfaces (exhibiting a roughness with scale-invariant self-similarity) that have recently been developed, in particular in Japan.[15] They can be either hydrophobic or hydrophilic. As we will describe in chapter 9, the surface roughness can control the degree of wettability (for a given surface chemistry) by enhancing the material's natural tendency. As the roughness increases, a hydrophilic substance becomes even more hydrophilic, while one that starts hydrophobic can become literally "super-hydrophobic" (see Figure 1.17).

Surface Treatments

Hydrophilic Surfaces Made Hydrophobic. We have seen that glass and silicon are high-energy solids that are wetted by all liquids (with the exception of mercury) because their critical surface tension γ_C is quite high (of the order of 150 mN/m). It is possible to lower γ_C by coating such solids with a hydrophobic molecular layer of the type $-(CH_2)-$ or $-(CF_2)-$. This trick can create surfaces with extremely low energies, mimicking Teflon (a fluorinated polymer). The parameter γ_C drops down to values of the order of 20 mN/m for hydrogenated coatings and 10 mN/m for fluorinated ones. Practically no

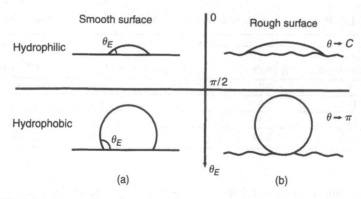

FIGURE 1.17. Controlling the wettability of a substrate through its roughness. Smooth surface (a); rough surface (b). Hydrophilic substrate becoming even more hydrophilic with a rough surface (top); hydrophobic substrate becoming "super-hydrophobic" (bottom).

liquid spreads on a fluorinated surface. On a hydrogenated surface, PDMS spreads totally if $\gamma_C > 21$ mN/m, and partially otherwise. Such surfaces turn out to be extremely interesting because they can be wetted totally by oils, while at the same time their low energy protects them against contamination effects and promotes stability over long periods of time. Two substances capable of altering the wetting properties of surfaces are octadecyltrichlorosilane (OTS), whose chemical formula is $Cl_3-Si-(CH_2)_{17}-CF_3$, and heptadecafluoro-1,1,2,2-tetrahydrodecyltrichlorosilane, whose formula is $Cl_3-Si-(CH_2)_2-(CF_2)_7-CF_3$.

Hydrophobic Surfaces Made Hydrophilic. Greenhouses are often covered with transparent plastic sheets. Morning dew condensing into fine droplets on the plastic scatters the light and robs flowers and plants of much needed sunlight. It is desirable to find a way to force water to spread into a continuous film, in other words, to "wet" the material. There are "plasma" treatments that can create hydrophilic groups on the surface of the plastic, thereby lowering γ_C.

The human cornea is extremely hydrophobic. Our tears "treat" the surface of the cornea by depositing hydrophilic proteins that stabilize the lachrymal film. Another interesting example is that of mushroom spores that can play havoc in rice plantations. Their destructive effect can be traced to their ability to alter the surface of the rice plant—normally very hydrophobic—by turning it hydrophilic, enabling the spores to readily attach themselves to it.

Plastics and molecular crystals generally have a low γ_C and, therefore, are poorly wettable by water. One technique to increase their wettability is to coat them with gold. However, it would be a mistake to believe that gold-coated plastic behaves like bulk gold. The liquid does interact with gold, but that does not mean that interactions with the plastic substrate are entirely masked. While a very thin liquid film "thinks" it sits on pure gold, a thick one will still "sense" the underlying substrate. This paradoxical situation leads to "pseudopartial" wetting, where the liquid covers the solid with an extremely thin film without truly spreading (the contact angle θ_E remains finite). This will be discussed in more detail in Section 4.2.3.

An Ideal Substrate: The Silicon Wafer

As stated earlier, silicon wafers of the type developed for the microelectronics industry are a popular choice of solid surface. In their natural state (that is to say, when stored in ordinary atmosphere), such wafers are coated with a thin layer of native oxide (SiO_2) about 14 Å thick. These surfaces bear a close resemblance to those of molten silica, particularly with regard to silanol groups (Si–OH). One of the primary advantages of these substrates is their planarity, flatness, and smoothness. X-ray studies reveal a residual roughness of no more than 5 Å. More detailed measurements suggest that the underlying silicon surface has a few atomic steps spaced by about a centimeter.

Cleaning. Before use, the surface must be carefully cleaned according to a process involving two steps.

The wafers are immersed for at least 30 minutes into an acid bath mixture of sulfuric acid and hydrogen peroxide (in the ratio of 70:30%) maintained at a temperature of 70°C. Next, they are rinsed in distilled water and dried in an oven at 100°C. At this point, they are exposed to UV radiation in an oxygen atmosphere. The ozone produced during this step breaks up any residual organic impurities that might remain on the surface.

A standard test for cleanliness consists in watching what happens to a water drop deposited on the surface of the silicon. Water wets "bare" silicon, whereas it does not spread in the presence of impurities. One can also exhale onto the surface and watch the result. If the surface is clean, water vapor coats the silicon in the form of a homogeneous film that evaporates uniformly; if not, a haze forms that disappears more slowly. This phenomenon has been studied in France; it is called "figure de souffle" (breath pattern).

Surface Treatments (Glass and Silicon).[16,17] Silicon surfaces belong in the high-energy (HE) category. Their critical tension γ_C for wetting exceeds 150 mN/m. As mentioned earlier, it is possible to modify this value by depositing a LE layer onto the solid. The parameter γ_C can be lowered to make the surface non-wetting relative to impurities and contaminants present in the ambient atmosphere or for the specific liquids chosen (i.e., PDMS's, alkanes, etc.).

Since γ_C depends essentially on the specific chemical groups on the surface, one can choose the type of coating to achieve a desired value of γ_C. A compact layer of methyl groups has a γ_C of 22 mN/m; a similarly compact

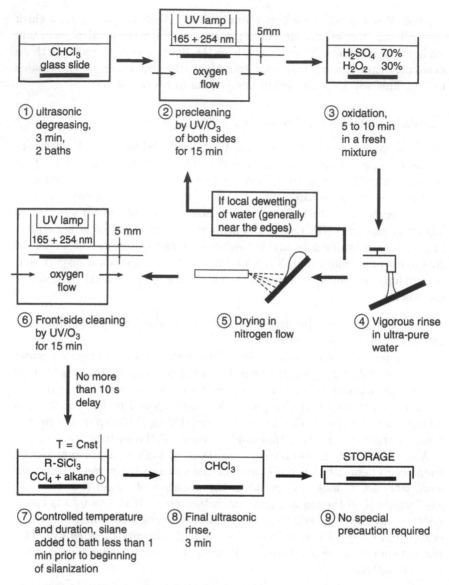

FIGURE 1.18. Flow diagram of the silanization process (courtesy J. B. Brzoska).

layer of fluorinated groups has a γ_C of only 6 mN/m. The reaction involved is known as *silanization*. A trichlorosilane group reacts chemically with the silanols on the surface, grafting one hydrophobic chain onto the substrate. The silanization process is illustrated in Figure 1.18.

One can obtain two different types of chemical surfaces.

1. *Silanized substrates* (OTS). This type itself splits into two subtypes depending on the compactness of the coating:

- "Fluffy" layer. When the coating is partial (meaning that the deposited layer does not form dense, continuous coating), γ_C comes out to be between 24 and 28 mN/m. In this case, wetting is total for silicone oils, although the surface is less sensitive to external conditions than the bare surface (which has a much higher γ_C).
- "Dense" layer. γ_C is equal to 21 ± 2 mN/m. The resultant substrates are then non-wetting as far as alkanes are concerned.

2. *"Teflon-like" surfaces.* Here γ_C is equal to 15 ± 2 mN/m. Neither silicone oils nor alkanes are then able to spread.

Glass

One routinely uses floated glass, which is somewhat less smooth than a silicon wafer and less pure as well because of the incorporation of inorganic substances during the floating process. Its advantage is to be much cheaper. Additionally, it is optically transparent. Its surface composition (silanols) is compatible with the various surface treatments described earlier, including cleaning and silanization.

Usual (Non-Ideal) Substrates

In practice, one deals with less smooth substrates displaying a wide variety of chemical properties: glass, plastics, ceramics, metals, and others. It is a challenge to make simple predictions concerning the interfacial energies of such materials. Nonetheless, a body of empirical knowledge is gradually building up, involving three main types of liquid/solid interactions:

- London–van der Waals forces, which have already been discussed;
- interactions between permanent dipoles;
- "acid/base"-type interactions.

These concepts have been developed by Good, Fowkes, Van Oss, and Chaudhury. The reader may want to consult a good text on the topic.[18] The effects of surface roughness will be discussed in detail in chapters 3 and 9.

1.2.4 Liquid Substrates: Neumann's Construction[19]

Consider a liquid A (oil) wetting a second liquid B (water). A and B are assumed to be immiscible. The surface of liquid B is no longer planar. Rather, it adjusts itself so as to minimize its surface energy. The contact angle is no longer given by Young's relation, derived earlier, but can be worked out by means of Neumann's construction (Figure 1.19). Both the horizontal and vertical components of the capillary forces must add up to zero. The construction is possible only if $S = \gamma_B - (\gamma_A + \gamma_{AB}) < 0$. If $S > 0$, liquid A wets liquid B and spreads totally on the free surface of B, as exemplified by PDMS on water.

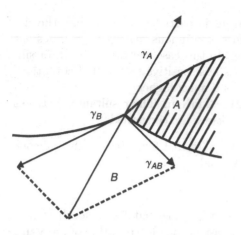

FIGURE 1.19. Neumann's construction.

The advantage of liquid substrates is that they are smooth on an atomic scale and chemically homogeneous. Furthermore, one can readily measure the three interfacial tensions γ_A, γ_B, and, γ_{AB}, a luxury not available with a solid substrate. The wettability can easily be adjusted by selecting liquids that are more or less polar.

The price to pay is that the substrate is not rigid and that it flows. When studying the dynamics of wetting, it becomes necessary to take into account the flows induced in the substrate as liquid A spreads or dewets (chapter 7).[20]

To avoid contamination effects, it is a good idea to select liquids with a low surface tension. A possible choice is the PDMS/fluoroalkylsiloxane pair (the latter is a fluorinated derivative of PDMS). These two liquids are immiscible. PDMS, which happens to be lighter, does not wet the fluorinated siloxane.

If the goal is to work with systems with very low viscosity, we recommend either water, which does not spread on CCl_4 or $CHCl_3$, or alkanes on water.

In chapter 2, we will describe a simple technique for determining the interfacial tension γ_{AB} by measuring the thickness of a floating lens.

Appendix: Minimal Surfaces – Euler-Lagrange Equations

Minimal surfaces can be calculated with the help of Laplace's formula [equation (1.6)]. Alternatively, one can minimize the surface, keeping the volume constant, by means of Euler-Lagrange's equations.

We proceed to demonstrate the method by rederiving the profile of a drop of initial radius R deposited on a fiber of radius b (see Figure 1.10).

The drop has a volume $\Omega = \frac{4}{3}\pi R^3$. We assume that the drop merges with the fiber with a contact angle equal to zero. At equilibrium, the surface energy of the drop is minimum, while the volume is constrained to a fixed value. Thus, we are faced with minimizing the function $G = \gamma A - \lambda \Omega$, where A is the surface area of the drop and the coefficient λ is a Lagrange multiplier, which has the dimension of a pressure. Indeed, we will show that λ is in fact the pressure difference between the inside of the drop and the outside medium.

The profile of the drop is described in terms of the distance $z(x)$ of its outer edge to the axis of the fiber. We have

$$G = 2\pi\gamma \int z\sqrt{1 + \dot{z}^2}\, dx - \lambda\pi \int (z^2 - b^2)\, dx \qquad (1.32)$$

where $\dot{z} = \frac{dz}{dx}$ and $2\pi z\sqrt{1 + \dot{z}^2}\, dx$ is a differential element of surface, taking into account the axial symmetry of the drop. We minimize the energy by using Euler-Lagrange's equations. If $G = \int f(z, \dot{z})\, dx$, the extremum of G satisfies the condition

$$-\frac{d}{dx}\left[\frac{\partial f}{\partial \dot{z}}\right] + \frac{\partial f}{\partial z} = 0. \qquad (1.33)$$

This can be integrated to

$$\dot{z}\frac{\partial f}{\partial \dot{z}} - f = cnst. \tag{1.34}$$

The first equation (1.33) corresponds to Newton's fundamental dynamics equation. It can be recast in the form

$$\gamma \left[\frac{-\ddot{z}}{(1 + \dot{z}^2)^{3/2}} + \frac{1}{z(1 + \dot{z}^2)^{1/2}} \right] = \lambda. \tag{1.35}$$

We have just rediscovered Laplace's formula $\gamma(\frac{1}{R} + \frac{1}{R'}) = \lambda$, which proves that $\lambda = \Delta p$.

The second equation (1.34) is equivalent to the principle of energy conservation. Its physical meaning here is the conservation of the force acting on a section of drop:

$$-\lambda\frac{z^2}{2} + \gamma\frac{z}{\sqrt{1 + \dot{z}^2}} = C. \tag{1.36}$$

The constant C is determined from the boundary conditions, namely, $\dot{z} = 0$ at $z = b$, which yields

$$-\lambda\frac{z^2 - b^2}{2} + \gamma\left(\frac{z}{\sqrt{1 + \dot{z}^2}} - b \right) = 0. \tag{1.37}$$

This last equation is identical with equation (1.15), which was written directly based on a force conservation argument.

The maximum radius L of the drop is obtained when $\dot{z} = 0$, which leads to $\frac{\lambda}{2} \cdot (L^2 - b^2) = \gamma \cdot (L - b)$. This last result simplifies to

$$L = \frac{2\gamma}{\lambda} - b. \tag{1.38}$$

References

[1] H. Bouasse, *Capillarité et phénomènes superficiels* (Capillarity and Surface Phenomena) (Paris: Delagrave, 1924).

[2] W. Wick, *Gouttes d'eau* (Water Drops) (Paris: Millefeuilles, 1998).

[3] J. Rowlinson and B. Widom, *Molecular Theory of Capillarity* (Oxford, U.K.: Oxford University Press, 1982).

[4] A. W. Adamson, *Physical Chemistry of Surfaces* (New York: John Wiley and Sons, 1990).

[5] Pierre Simon de Laplace, *Oeuvres complètes de Laplace, t IV, Supplément au livre X du traité de la mécanique céleste* (Complete Works of Laplace, tome 4, Supplement to Book 10 of the Treatise on Celestial Mechanics), p. 394. Also *2ème supplément au livre X* (2nd Supplement to Book 10), p. 419, ch. 1.

[6] J. Plateau, *Statique expérimentale et théorique des liquides soumis aux seules forces moléculaires* (Experimental and Theoretical Equilibrium State of Liquids Subjected to Molecular Forces Only) (Paris: Gauthiers-Villars, 1873).

[7] Lord Rayleigh, *Scientific Papers*, (Cambridge: Cambridge University Press, 1899): *Proc. London Math. Soc.* **10**, 4 (1878); *Proc. R. Soc. London* **29**, 71 (1879); *Philos. Mag.* **34**, 145 (1873).

[8] H. Poincaré, *Capillarité* (Capillarity) (Paris: G. Carré, 1895).

[9] T. Young, *Philos. Trans. Soc. London* **95**, 65 (1805).

[10] J. Fox and W. Zisman, *J. Colloid Interface Sci.* **5**, 514 (1950).

[11] W. Zisman, Contact Angle Wettability and Adhesion, in *Chemical Series*, 43 (Washington, D.C., 1964).

[12] P. G. de Gennes, *Rev. Mod. Phys.* **57**, 827 (1985).

[13] J. F. Joanny, *Doctoral Thesis, University of Paris* (1985).

[14] L. Monnerie and G. Champetier, *Macromolécules* (Macromolecules) (Paris: Masson, 1969).

[15] T. Onda, S. Shibuichi, N. Satoh, and K. Tsujii, *Langmuir* **12**, 2125 (1996).

[16] J. Sagiv, *J. Am. Chem. Soc.*, **102**, 92 (1980).

[17] J. B. Brzoska, N. Shahizadeh, and F. Rondelez, *Nature* **360**, 24 (1992).

[18] C. J. Van Oss, *Interfacial Forces in Aqueous Media* (New York: Marcel Dekker, 1994).

[19] I. Langmuir, *J. Chem. Phys.* **1**, 756 (1933).

[20] J. F. Joanny, *Physico Chemical Hydrodynamics* **9**, 189 (1987); A. M. Cazabat, *J. Colloid Interface Sci.* **133**, 452 (1989).

2
Capillarity and Gravity

Liquids display rather peculiar properties. They have the ability to defeat gravity and create capillary bridges (see Figure 2.1), move up inclined planes, or rise in very small capillary tubes.[1] Moreover, drops may lose their spherical shape under the influence of gravity.

2.1 The Capillary Length κ^{-1}

There exists a particular length, denoted κ^{-1}, beyond which gravity becomes important. It is referred to as the *capillary length*. It can be estimated by comparing the Laplace pressure γ/κ^{-1} to the hydrostatic pressure $\rho g \kappa^{-1}$ at a depth κ^{-1} in a liquid of density ρ submitted to earth's gravity $g = 9.8$ m/s^2. Equating these two pressures defines the capillary length:

$$\kappa^{-1} = \sqrt{\gamma/\rho g} \tag{2.1}$$

The distance κ^{-1} is generally of the order of few mm (even for mercury, for which both γ and ρ are large). If one wants to increase κ^{-1} in a liquid by a factor 10 to 1,000, it is necessary to work in a microgravity environment or, more simply, to replace air by a non-miscible liquid whose density is similar to that of the original liquid.

Gravity is negligible for sizes $r < \kappa^{-1}$. When this condition is met, it is as though the liquid is in a zero-gravity environment and capillary effects

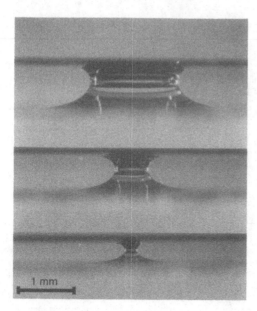

FIGURE 2.1. Liquid bath rising to form a capillary bridge (From "Nucleation Radius and Growth of a Liquid Meniscus," by G. Debregeas and F. Brochard-Wyart. In *Journal of Colloid and Interface Science, 190*, p. 134 (1997), © 2001 by Academic Press. Reproduced by permission.)

dominate. The opposite case, when $r > \kappa^{-1}$, is referred to as the "gravity" regime.

The distance κ^{-1} can also be thought of as a *screening length*. If one perturbs an initially horizontal liquid surface by placing on it a small floating object (Figure 2.2a), the perturbation induced on the surface dies out in a distance κ^{-1}.

It is convenient to consider the one-dimensional situation illustrated in Figure 2.2b. Here, the perturbing object is a solid vertical wall located at $x = 0$. In the vicinity of the wall, the surface of a liquid is no longer flat but, instead, rises to a height $z(x)$. Assume that the height z remains small (which is always true far enough from the wall, when $\kappa x > 1$). The local

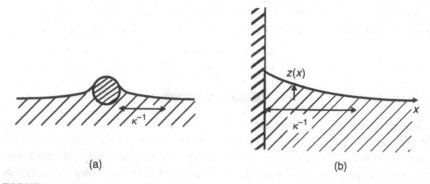

FIGURE 2.2. A small floating object (a) or a wall (b) perturbs the surface of the liquid over a distance κ^{-1} called the "capillary length."

curvature is then simply $-\frac{\partial^2 z}{\partial x^2}$, and Laplace's equation dictates that the pressure immediately under the surface be

$$p_A = p_{atm} - \gamma \frac{\partial^2 z}{\partial x^2} \qquad (2.2)$$

where p_{atm} is the atmospheric pressure. At the same time, the pressure must also conform to the laws of hydrostatics, that is to say,

$$p_A = p_{atm} - \rho g z. \qquad (2.3)$$

Equating these last two equations yields

$$\gamma \frac{\partial^2 z}{\partial x^2} = \rho g z \qquad (2.4)$$

or, equivalently,

$$\frac{\partial^2 z}{\partial x^2} = \kappa^2 z. \qquad (2.5)$$

The solutions to this last equation are of the form $z = z_0 \exp(\pm \kappa x)$. In the problem at hand, we are restricted to $z \to 0$ when $x \to \infty$. Therefore, we retain only the exponentially decaying solution:

$$z = z_0 \cdot \exp(-\kappa x) \qquad (2.6)$$

Conclusion. Surface perturbations decay *exponentially* with distance with a characteristic length κ^{-1}, which is the capillary length. In the immediate vicinity of the wall, the exact solution is more complicated because the height z is no longer necessarily small. This case will be discussed shortly (section 2.3).

2.2 Drops and Puddles in the Partial Wetting Regime

2.2.1 The Shape of Drops

Imagine that we place drops of increasing sizes on a piece of silanized glass or on a horizontal sheet of plastic. The largest drops tend to flatten under the influence of gravity (Figure 2.3).

We have seen in the previous chapter that, when the spreading parameter S is negative (partial wetting regime), a drop of liquid deposited on a horizontal substrate exhibits a contact angle θ_E determined by Young's law. The shape of the drop will therefore change from perfectly spherical to almost completely flat depending on whether its radius R is small or large compared to the capillary length κ^{-1}.[2]

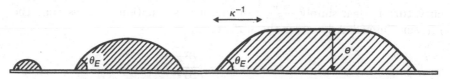

FIGURE 2.3. Water drops of increasing size on a sheet of plastic. Gravity causes the largest drops to flatten.

2.2.2 Droplets $(R \ll \kappa^{-1})$

For small drops of radius less than κ^{-1}, the capillary forces are the only ones to come into play. In accordance with Laplace's equation, their curvature must be constant. Therefore, a drop deposited on a horizontal surface takes on the shape of a spherical cap whose edges intersect the substrate at angle θ_E. Measuring that angle enables us to determine the spreading parameter (negative in the present case) through the expression $S = \gamma \cdot (\cos \theta_E - 1)$.

Drops Deposited on Dirty Surfaces. We have implicitly assumed an ideal surface. On a real surface, the contact angle of a drop is often slightly dependent on the preparation conditions. Its value lies between two limits θ_A (larger) and θ_R (smaller). The *hysteresis* of the contact angle, determined via the force $\delta = \gamma \cdot (\cos \theta_R - \cos \theta_A)$, will be discussed in chapter 3. The difference $\theta_A - \theta_R$ is a measure of the state of cleanliness and roughness of a surface. It is used as a test in the automobile industry to ensure that surfaces are perfectly clean before applying paint. $\theta_A - \theta_R$ must be sufficiently small for good adhesion.

2.2.3 Heavy Drops $(R \gg \kappa^{-1})$

For large drops whose radius exceeds κ^{-1}, gravitational effects dominate. A drop is flattened by gravity. At equilibrium, it takes on the shape of a liquid pancake of thickness e. The value of e can be calculated by expressing the equilibrium of the horizontal forces acting on a portion of the liquid.

The forces involved are shown in Figure 2.4. They are of two types:

1. Surface forces, which add up to $\gamma_{SO} - (\gamma + \gamma_{SL})$,
2. Hydrostatic pressure $\tilde{P}$, integrated over the entire thickness of the liquid, which amounts to $\tilde{P} = \int_0^e \rho g(e - \tilde{z}) \, d\tilde{z} = \frac{1}{2} \rho g e^2$.

The equilibrium of forces per unit length can be expressed by the equation

$$\frac{1}{2} \rho g e^2 + \gamma_{SO} - (\gamma + \gamma_{SL}) = 0 \tag{2.7}$$

which leads to:

$$S = -\frac{1}{2} \rho g e^2. \tag{2.8}$$

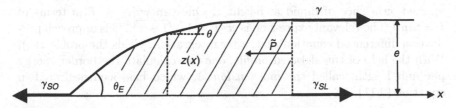

FIGURE 2.4. Equilibrium of the forces (per unit length of the line of contact) acting on the edge of a puddle. $\tilde{P} = \rho g e^2 / 2$ is the hydrostatic pressure.

Young's law, which describes the equilibrium of forces acting on the line of contact, implies that $\gamma_{SO} - (\gamma \cos\theta_E + \gamma_{SL}) = 0$. Therefore,

$$\gamma \cdot (1 - \cos\theta_E) = \frac{1}{2}\rho g e^2. \tag{2.9}$$

From the preceding equation, the thickness e of a puddle can be recast in terms of the capillary length:

$$e = 2\kappa^{-1}\sin\left(\frac{\theta_E}{2}\right). \tag{2.10}$$

When $\theta_E \ll 1$, the thickness simplifies to $e = \kappa^{-1}\theta_E$.

Alternatively, the thickness e can be calculated by minimizing the energy F_g of the puddle. If the surface area A of the drop is very large ($\sqrt{A} \gg \kappa^{-1}$), it is legitimate to neglect the energy associated with the edges. The energy of the drop can then be written as

$$F_g = -SA + \frac{1}{2}\rho g e^2 A. \tag{2.11}$$

The procedure consists in minimizing the energy F_g while keeping the volume $\Omega = Ae$ constant, which leads straight back to equation (2.8).

> **The Housewife Problem:** A bucket containing 6 liters of water is emptied onto the ground. Calculate the wet surface area A for $\theta_E = 180°$ and for $\theta_E = 1°$. (Answers: 1 m²; 120 m².)

Detailed Profile

To calculate the profile of a drop near its edge, one can still express the equilibrium of the horizontal forces acting on a portion of the drop near the boundary (Figure 2.4):

$$\gamma_{SO} + \tilde{P} = \gamma\cos\theta + \gamma_{SL} \tag{2.12}$$

where $\tilde{P} = \int_0^z \rho g(e - \tilde{z})\, d\tilde{z} = \rho g(ez - \frac{z^2}{2})$ is the hydrostatic contribution exerted on a slice of liquid at height z. One can write $\cos\theta$ in terms of $\dot{z} = \tan\theta$. The relevant expression is $\cos\theta = 1/\sqrt{1+\dot{z}^2}$. This approach produces a differential equation that, after integration, yields the profile $z(x)$. With the help of this detailed profile, one can calculate the border energy per unit length, called the *line tension* $\Im$, which has been neglected in equation (2.11).

2.2.4   Experimental Techniques for Characterizing Drops

The goal is to measure all the relevant parameters that characterize the process of spreading, that is to say, the time evolution of a drop and its final steady state. The techniques used generally involve an optical setup and a camera to record the process continuously in real time. We will now review several basic methods used to determine specific parameters, such as the contact angle, the thickness of a liquid film, and the radius of a drop.

Measuring the Contact Angle of a Liquid on a Solid

Drop Acting as a "Mirror" ($1° < \theta < 45°$). This measurement is based on optical reflectometry.[3] The drop (with a typical diameter d of 5 mm) is used as a convex mirror whose edges contact the horizontal support with an angle θ to be determined. When illuminated by a collimated laser beam (diameter of the order of 5 mm) normally incident on the substrate, the drop reflects a cone of light whose total angular divergence is 4θ (Figure 2.5a). A simple method is to place a horizontal screen (translucent glass or wax paper) at a height h above the substrate and to measure the diameter D

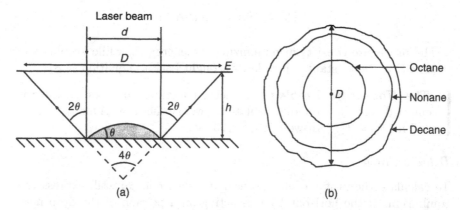

(a)   (b)

FIGURE 2.5. (a) Drop illuminated by a collimated light beam. The drop reflects a cone of light that can be observed on a translucent screen E placed at a height h above the substrate. (b) Record of the diameter of a laser beam reflected by a drop of alkane deposited on silanized glass.

TABLE 2.1.

Substance	d (mm)	γ (mN/m)
Octane	5	21.8
Nonane	3.5	22.9
Decane	3	23.9
Heptane	Spreads	20.2

of the reflected cone. The contact angle is then given by the geometrical expression

$$\tan(2\theta) = \frac{D-d}{2h}. \tag{2.13}$$

The accuracy on the measurement of θ is about 0.1°. The angle deduced from the average value of the diameter D is not a local value (as would be the case for more traditional measurements using a goniometer). Rather, its value is integrated over the entire periphery of the drop (Figure 2.5), averaging out certain local irregularities of the contact line.

> **Problem:** We wish to characterize the surface treatment of a wafer of silanized glass (Figure 2.5).
>
> The experimental data available are $D/2 = 2.1$ cm for octane, 3.8 cm for nonane, and 4.5 cm for decane. The screen is placed at a height $h = 12$ cm. Additional known data are listed in Table 2.1.
>
> Determine the contact angle for a drop of each substance. Using the curve describing the relation $\cos\theta = f(\gamma)$, deduce the critical surface tension γ_C of the treated surface.

Projection. For angles θ_E greater than 45°, the measurement relies on an optical imaging technique (the previous setup is inherently limited to angles less than 45°). The drop is placed between an intense light source and a converging lens of focal lens F. The projected image is displayed on a screen located at a large distance L. Since the silicon support is reflecting, one obtains on the screen a symmetrical figure magnified in the ratio L/d and exhibiting the same contact angle as the original one (Figure 2.6b). The measurement accuracy is of the order of 2°. Although the present technique is usable for any value of θ, the previous one remains preferred whenever *theta* is less than 45°.

Interference Contrast Microscopy in Reflection (RICM)

Consider a drop of a transparent liquid, characterized by a low contact angle (a few degrees), deposited on a smooth and reflecting substrate. When such a drop is observed under a microscope, circular and concentric black fringes appear. These fringes are due to interference effects between the

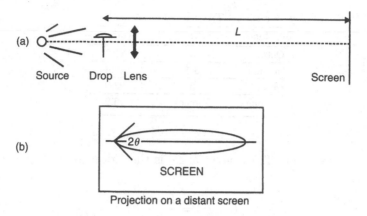

FIGURE 2.6. Optical projection method.

beam reflected by the surface of the drop and that reflected by the underlying portion of substrate covered by the drop. Black fringes correspond to destructive interference and occur only for specific thicknesses of the liquid. They can be thought of as contour lines producing a topographic map of the drop's profile. The fringe pattern therefore conveys important information about the contact angle, the radius of the spherical cap, and the volume of the drop.

The basic principle behind the technique is illustrated in Figure 2.7. An incident optical beam of intensity I_0 is partially reflected by the liquid/air interface as a beam of intensity I_1, while the transmitted part of the incident beam is reflected by the substrate immersed in the liquid as a beam of intensity I_2. These two beams interfere with each other and give rise to fringes of constant thickness.

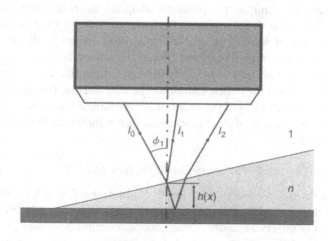

FIGURE 2.7. Formation of fringes of equal thickness in a liquid wedge due to interference between reflected beams I_1 and I_2.

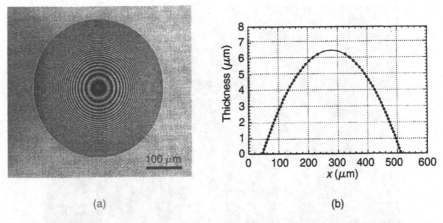

(a)   (b)

FIGURE 2.8. Image of a microdrop of PDMS on a silanized silicon wafer. The image was obtained by intereference microscopy in reflection. From the interference fringe pattern (a), one can reconstruct the drop's profile (b) (courtesy R. Fondecave).[4]

Figure 2.8a shows a digitized image of a drop of PDMS on a silanized silicon substrate. Manipulating such an image with a pattern analysis algorithm makes it possible to reconstruct the profile of the drop.

By associating each fringe with its corresponding height above the substrate, the precise geometrical characteristics of the drop, including its contact angle, can be extracted (Figure 2.28b).[4] Note the high precision of this technique for measuring θ (in this case, $\theta = 20.4° \pm 0.1°$).

Distortion of the Image of a Grid by a Liquid Wedge (the "Grid Technique")

The "grid technique" for measuring contact angles is particularly well suited for time-dependent measurements, although it is equally applicable in steady-state situations. It requires transparent substrates and liquids. The technique relies on the distortion of a grid viewed through the edges of the drop. As an example, Figure 2.9 shows the image obtained in a dewetting experiment. The grid is distorted in the vicinity of the line of contact encircling a hole in a film (the same effect would be seen at the edge of a drop deposited on a substrate).[5]

Measuring the Thickness of a Drop

This method (devised by F. Rondelez) is based on manipulating a thin metal needle mounted on a vertical translation stage to bring it in contact first with the liquid/air interface and second with liquid/solid interface, as depicted in Figure 2.10. The thickness of the film (or of the drop) is deduced from the difference between these two vertical positions read on a micrometer. When reaching contact with the liquid, a meniscus forms instantly, which is easily detected with the naked eye. The suddenness of

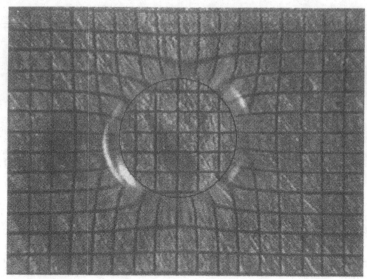

1 mm = 0.75 cm

FIGURE 2.9. Optical distortion of a grid pattern on a substrate. The distortion is caused here by a hole in a film (courtesy C. Andriue). (From "A New Method for Contact Angle Measurements of Sessile Drops," by C. Allain, D. Ausseré, and F. Rondelez. In *Journal of Colloid and Interface Science, 107*, p. 5 (1995), © 2001 by Academic Press. Reproduced by permission.)

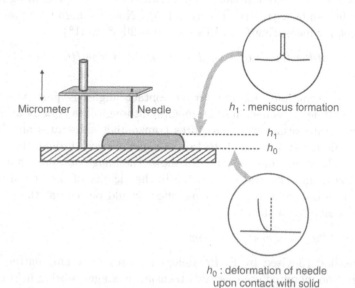

h_0 : deformation of needle upon contact with solid

FIGURE 2.10. Mechanical measurement of the thickness of a puddle. The height h_1 is recorded when the needle contacts the liquid, which is indicated by the formation of a meniscus. Contact with the solid substrate at height h_0 is detected when the needle bends.[6]

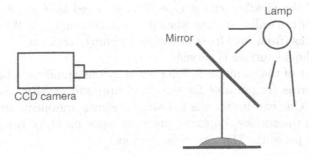

FIGURE 2.11. Apparatus for measuring the radius of a drop as a function of time.

the reaction at the upper contact even provides an excellent measurement accuracy, which is limited only by the resolution of the micrometer, or about 10 μm. Contact with the solid surface is signaled by the bending of the needle as its tip touches the solid. This bending is detected through the sudden and drastic change in the diffraction pattern of a collimated laser beam skimming over the straight needle.[6] Accuracy is again 10 μm.

This technique is able to determine liquid film thicknesses over a wide range (preferably more than 100 μm). It works equally well for opaque and transparent liquids. In the case of fluids (viscosity η of the order of one mPa-s), the relaxation time following contact is quite short and measurements can be repeated in rapid succession, making it possible to follow the dynamical evolution of a drop in the process of spreading.

Measuring the Radius of a Drop

A simple technique for measuring the radius of a drop is to image it on a screen with a video camera equipped with a zoom lens (using a magnification of typically 20×) (Figure 2.11). Lengths can be measured with an accuracy of about 100 μm.

For surfaces exhibiting wetting hysteresis, the advancing angle θ_A and the receding angle θ_R, to be studied in chapter 3, are measured using the optical techniques just described. The liquid is deposited on a surface placed within an enclosed sample chamber (to curb evaporation) by means of a syringe. The angle θ_A is measured as the liquid is inserted; the angle θ_R is measured as the liquid is withdrawn.

2.3   Menisci

2.3.1   Characteristic Size

A liquid is normally contained in a solid vessel with vertical walls. The surface of the liquid is horizontal because of gravity, except near the walls where Young's relation [equation (1.23)] induces a distortion. When the liquid is of the mostly wetting type (characterized by $\theta_E < 90°$), it rises slightly near the walls, whereas when it is non-wetting ($\theta_E > 90°$), it drops. The meniscus (from the Greek *mêniskos*, meaning "crescent") is the region where the liquid surface is curved.

The shape of the meniscus is determined by the equilibrium between the capillary forces (responsible for the curvature) and gravity forces (which oppose it). One can invoke the following pressure argument: Immediately underneath the surface, Laplace's pressure [equation (1.6)] is equal to the hydrostatic pressure. This can be written as

$$P_0 + \frac{\gamma}{R(z)} = P_0 - \rho g z \tag{2.14}$$

where z is the height of the surface above the level of the bath, P_0 is the outer pressure, and $R^{-1}(z)$ is the curvature at a particular point. This expression shows that for $z > 0$ (ascending meniscus), the curvature is negative, which does indeed correspond to a liquid under suction. Since the height z varies from point to point along the interface from 0 (the altitude of the bath proper) to h at the wall, so does the radius of curvature of the interface. As a result, the meniscus is not shaped like a circle. Equation (2.14) simplifies to

$$-Rz = \kappa^{-2} \tag{2.15}$$

where, again, we run into the capillary length $\kappa^{-1} = \sqrt{\gamma/\rho g}$. Ordinary experience indicates that the typical dimension of the meniscus is of the order of a millimeter. Note, however, that as told before the capillary length can be much larger if the densities of the liquid and the fluid above it are comparable. In this case, the density ρ of the liquid, which enters the expression of the capillary length, should be replaced by the difference $\Delta\rho$ between the densities of the two fluids. Likewise, in a reduced gravity environment ($g \to 0$), the capillary length diverges, in which case the meniscus extends over the entire surface of the bath, as illustrated in Figure 2.12. Some people believe that without gravity a wetting liquid would simply float out of a glass; in actuality, it experiences an underpressure because of surface tension and has no reason whatsoever to escape.

As a general rule, the characteristic size of a meniscus is either the capillary length κ^{-1} or the size l of the vessel itself, whichever is smaller.

FIGURE 2.12. Meniscus in a glass of water in reduced gravity.

2.3.2   Shape of a Meniscus Facing a Vertical Plate

We study here the meniscus of a liquid facing a vertical plate, shown in Figure 2.13. We now return to the general equation (2.15). Here one of the curvature radius is infinite. The curvature $C = 1/R$ can be expressed as

$$\frac{1}{R} = -\frac{d\theta}{ds} \tag{2.16}$$

where θ is the angle between the tangent to the curve at a particular point and the vertical direction, and s is the curvilinear coordinate.

Switching to a Cartesian coordinate system in the (x, y) plane (Figure 2.13), equation (2.16) translates to

$$\frac{1}{R} = -\frac{\dfrac{d^2 z}{dx^2}}{\left[1 + \left(\dfrac{dz}{dx}\right)^2\right]^{3/2}}. \tag{2.17}$$

Equation (2.17) is a second-order differential equation for the profile $z(x)$ of the meniscus, subject to the appropriate boundary conditions. A first

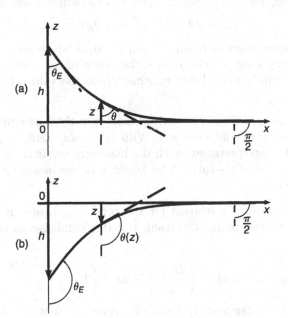

FIGURE 2.13. Menisci for a liquid that is (a) attracted to or (b) repelled by a vertical wall. Case (a) corresponds to $\theta_E < \pi/2$ and case (b) to non-wetting ($\theta_E > \pi/2$). The profile $z(x)$ goes exponentially to zero when $\kappa x > 1$ [equation (2.6)].

integration, together with the boundary condition $\frac{dz}{dx} = z = 0$ when $x \to \infty$ (where the profile merges with the horizontal surface of the bath), yields

$$\frac{1}{\left[1 + \left(\dfrac{dz}{dx}\right)^2\right]^{1/2}} = 1 - \frac{z^2}{2\kappa^{-2}}. \tag{2.18}$$

The same result can be derived independently without any calculus by expressing the equilibrium of the forces acting a portion of the meniscus extending between x and infinity. The pertinent contributions are the capillary forces and the forces associated with the hydrostatic pressure P_0 outside the liquid and $P_z = P_0 - \rho g \tilde{z}$ inside the liquid. The resultant pressure force $\tilde{P} = \int_0^z \rho g \tilde{z} \, d\tilde{z} = \frac{1}{2}\rho g z^2$ exerts itself in the direction $x < 0$. The argument is similar to the one used in connection with Figure 2.3, even though the substrate is now vertical. In *horizontal* projection, the equilibrium can be written as

$$\gamma \sin\theta + \frac{1}{2}\rho g z^2 = \gamma \tag{2.19}$$

which is but a slightly disguised version of Equation (2.18).

Equation (2.19) allows us to calculate the height h to which the meniscus rises. At $z = h$, Young's condition gives $\theta = \theta_E$, which leads to

$$h = \sqrt{2} \cdot \kappa^{-1} \cdot (1 - \sin\theta_E)^{1/2}. \tag{2.20}$$

The maximum height is reached when $\theta = 0$, at which point $h = \sqrt{2}\kappa^{-1}$. When observing water on clean glass, the height of the meniscus turns out to be 4 mm, which gives direct information on the value of the capillary length.

Remark. Equation 2.20 can be derived more directly from equations (2.14) and (2.16) leading to $\gamma \, d\theta/ds = \rho g z$. With $ds = -dz/\cos\theta$, we get $\rho g z \, dz = -\gamma \cos\theta \, d\theta$. By interpretation, with the boundary condition $\theta = \frac{\pi}{2}$ for $z = 0$, we get $\frac{1}{2}\rho g z^2 = \gamma(1 - \sin\theta)$. The height h of the meniscus is the value of z for $\theta = \theta_E$.

To obtain an explicit relation for the profile $z(x)$, one must integrate equation (2.18) once more. The result is rather cumbersome and uninspiring:

$$x - x_0 = \kappa^{-1} \cosh^{-1}\left(\frac{2\kappa^{-1}}{z}\right) - 2\kappa^{-1}\left(1 - \frac{z^2}{4\kappa^{-2}}\right)^{1/2} \tag{2.21}$$

where x_0 is the distance such that (2.21) gives $z = h$ at $x = 0$ (i.e., at the wall).

Finally, the equilibrium of forces over the entire meniscus in *vertical* projection is

$$\gamma \cos\theta_E = \int \rho g z(x) \, dx. \tag{2.22}$$

In the partial wetting regime, $\gamma \cos \theta_E = \gamma_{SO} - \gamma_{SL}$. This shows that $\gamma_{SO} -$ γ_{SL} is the force per unit length that supports the weight of the meniscus $W_{meniscus}$ (also expressed per unit length). In the total wetting regime, the meniscus connects smoothly with the wall with an angle $\theta_E = 0$. The weight of the meniscus is therefore $W_{meniscus} = \gamma < \gamma_{SO} - \gamma_{SL}$. The total force $\gamma_{SO} - \gamma_{SL}$ offsets not only the weight of the meniscus (γ), but also that of the wetting film rising high above the liquid bath (the existence of this film will be discussed in more detail later). In summary,

In total wetting regime:		In partial wetting regime:	
$\left.\begin{array}{l}\gamma = W_{meniscus} \\ S = W_{film}\end{array}\right\}$	(2.23)	$\left.\begin{array}{l}\gamma_{SO} - \gamma_{SL} = W_{meniscus} \\ \text{No Film}\end{array}\right\}$	(2.24)

2.3.3   Meniscus on a Vertical Fiber

We have already touched on this case in chapter 1 [section (1.6)] as an example of a surface with zero curvature. We can carry the analysis further by including the range of the perturbation created on the liquid surface far from the fiber, where gravity is no longer negligible. If the solid is a capillary rod of radius b less than the capillary length κ^{-1}, experiments show that the height of the meniscus is clearly smaller than that predicted by equation (2.20), because it must join with the surface of the fiber, which has a high curvature.

As in the case of a planar wall, the profile of the interface can be calculated by equating the Laplace and hydrostatic pressures. However, because of the additional curvature of the fiber, equation (2.14) must be modified to

$$P_0 + \gamma \left(\frac{1}{R_1} + \frac{1}{R_2} \right) = P_0 - \rho g z \qquad (2.25)$$

where R_1 and R_2 are the two principal radii of curvature at any point on the interface. We could, of course, integrate equation (2.25) numerically. Instead, we will try to give a more physical picture of what goes on. Near the line of contact, curvatures are of the order of b^{-1}. In other words, they are much more dominant here than in the planar case, when they were of the order of κ. Each curvature term here far exceeds the hydrostatic term in equation (2.25), to the point that gravity can be neglected altogether. This approximation is valid in the vicinity of the fiber only, at distances $r < \kappa^{-1}$. This constitutes an a posteriori justification of the argument used in section 1.1.6. The equation of the meniscus is then that of a surface with zero curvature

$$\frac{1}{R_1} + \frac{1}{R_2} = 0. \qquad (2.26)$$

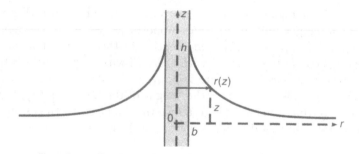

FIGURE 2.14. Meniscus on a fiber.

The meniscus adopts the form of a catenary curve, as discussed in chapter 1. In the coordinate system defined in Figure 2.14, the profile $r(z)$ reads

$$r(z) = b \cosh\left(\frac{z - h}{b}\right) \tag{2.27}$$

where the liquid has been assumed to be wetting ($\theta_E = 0$) to simplify the algebra. The integration constant h gives the height of the meniscus, which can be evaluated by noting that the profile given by equation (2.27) cannot merge with the vessel at infinity. Gravity actually prevents the meniscus from extending indefinitely. By constraining the lateral dimension not to exceed the capillary length κ^{-1}, one obtains:

$$h \approx b \ln\left(\frac{2\kappa^{-1}}{b}\right) \tag{2.28}$$

which is just a few times the radius of the fiber (for $b = 10\ \mu\mathrm{m}$ and $\kappa^{-1} = 1.5$ mm, the result is $h = 60\ \mu\mathrm{m}$). The lateral extent κ^{-1} of the meniscus is thus very much larger than its height.[7] This is in sharp contrast to the planar case, for which these two quantities are of the same order of magnitude. Note also that at distances $r > \kappa^{-1}$, the fiber creates a surface perturbation described by equation (2.4) generalized to two dimensions:

$$\gamma\left(\frac{\partial^2 z}{\partial z^2} + \frac{\partial^2 z}{\partial y^2}\right) = \rho g z.$$

One application of these principles is in the fabrication of microtips of the type used in atomic force microscopes (AFM). A fiber can be dipped in an acid bath to etch the material away.[8] The radius of the fiber diminishes as time progresses, which causes the height of the meniscus to drop in accordance with equation (2.28). The process goes on until all that is left is a sharp tip resting on the surface of the acid bath (Figure 2.15).

Figure 2.16 shows such a tip made by placing a glass fiber at the interface between hydrofluoric acid and cyclohexane. The initial fiber radius is 1 mm, which is much less than the capillary length (of the order of 1 cm for this particular system).

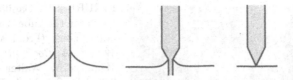

FIGURE 2.15. Fabrication of an AFM tip, by partially immersing a fiber in an etching solution.

FIGURE 2.16. Glass fiber placed at the interface between hydrofluoric acid and oil. After a few minutes, a tip forms where the meniscus used to be. (From "Meniscus Shape on Small Diameter Fibers," by K. M. Takahashi. In *Journal of Colloid and Interface Science, 134*, p. 181 (1990), © 2001 by Academic Press. Reproduced by permission.)

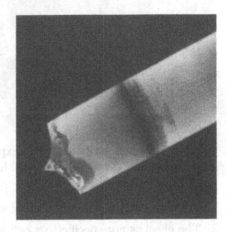

2.4 Capillary Rise in Tubes: Jurin's Law

2.4.1 Historical Background

When a narrow tube is brought is contact with a mostly wetting liquid, some of the liquid rises inside the tube (Figure 2.17). That is one of the most well-known and vivid manifestations of capillarity. Indeed, it is the very foundation of the entire field and it deserves a short historical digression. Leonardo da Vinci (1452–1519) was the first to observe the phenomenon, which he recorded in his notes. He concluded (rather boldly) that mountain springs are due to a network of fine capillaries capable of lifting water. Jacques Rohault (1620–1675) interpreted the capillary rise of liquids as an inability of air to properly circulate in the narrow tube, thereby creating a vacuum. The astronomer Geminiano Montanari (1633–1687) compared the rise of a liquid in a tube to that of sap in a plant. Giovanni Borelli (1608–1679), better known for his observations of Jupiter's moons and his research on the locomotion of animals (notably flees), demonstrated in 1670 that the height reached by a liquid is inversely proportional to the radius of the tube.

But the real star in the story, a man unjustly forgotten today, is Francis Hauksbee (1713), who was the first to study the phenomenon in a

FIGURE 2.17. Capillary rise. The narrower the tube, the higher the liquid rises. (From a *A Drop of Water: A Book of Science and Wonder*, by Walter Wick. Published by Scholastic Press, a division of Scholastic Inc. Photograph © 1997 by Walter Wick. Reproduced by permission.)

systematic way. A series of careful experiments (often carried out with colored liquids and glass tubes) allowed him to draw the following conclusions:

1. The rise of the liquid occurs in air just as well as in vacuum (refuting Rohault's theory that the lack of air is responsible),

2. The effect is not specific to a cylindrical geometry. The liquid rises just as well between two parallel plates close to each other, but the height reached is then half that attained in a tube of diameter equal to the separation between the plates.

3. The effect also occurs with other solids (marble, brass) and other liquids (alcohol, essence of turpentine, "common" oil).

4. Finally, the height does not depend on the thickness of the tube's walls. When Hauksbee tried two tubes with identical inner diameters but outer diameters differing by a factor of ten, he observed that the liquid rose to the same height.

Hauksbee was a colleague of Isaac Newton. Newton described these various experiments in his famous treatise "Opticks" (unfortunately, he neglected to mention Hauksbee's name). In 1712, the mathematician Brook Taylor, the father of Taylor's expansions in mathematics, also conducted an experiment on capillary effects between two nearly parallel glass plates forming a small angle wedge. As he was looking from the side, he was amazed to see that the boundary of the liquid seemed to have the shape of a hyperbola (the height increasing as the spacing between the plates decreases). Hauksbee himself confirmed these observations with precise measurements. By some dark twist of fate, none of these people ended up with their name associated with the law of capillary rise. Instead, that honor went to the English physiologist James Jurin (1684–1750), who in 1718 independently confirmed that the height reached by a liquid is in inverse proportion to the diameter of the tube.

2.4.2   The Law of Capillary Rise

A full century was to pass before Laplace gained a complete understanding of the phenomenon in 1806. He realized that the liquid rises if the dry tube has a surface energy γ_{SO} greater than the surface energy γ_{SL} of the same tube when wet. Just as we have defined a spreading parameter $S = \gamma_{SO} - (\gamma_{SL} + \gamma)$, one can likewise define an imbibition (or impregnation) parameter I by:

$$I = \gamma_{SO} - \gamma_{SL}. \tag{2.29}$$

When I is positive, the system lowers its energy if a wet surface is substituted for a dry one. In this case, one observes a rise of the liquid (or imbibition if one deals with porous materials). When I is negative, on the other hand, the liquid level drops in the tube (capillary descent), since the energy is lower for a dry surface than for a wet one. Young's relation ($\gamma_{SO} - \gamma_{SL} = \gamma \cos \theta_E$) shows that the impregnation criterion ($I > 0$) can equivalently be written as $\theta_E < 90°$. As we have seen in chapter 1, a liquid that satisfies this condition is often referred to as a *mostly wetting liquid*. It might be more accurate to talk of an *impregnating liquid* instead.

Interestingly, the impregnation criterion ($I > 0$) is much less restrictive than the spreading criterion ($S > 0$) since $I = S + \gamma$. This explains why we see most liquids spontaneously soak sponges and other porous materials, whereas complete spreading is far less prevalent. In the first case, one surface is replaced by another, while in the second it is replaced by two others, which is a priori less favorable.

The energy E of a liquid column can be written in terms of its height h and the radius R of the capillary:

$$E = -2\pi R h I + \frac{1}{2}\pi R^2 h^2 \rho g \tag{2.30}$$

where the first term is the gain in surface energy and the second is the cost in terms of gravitational potential energy. This expression ignores the details of the meniscus and is therefore valid in the limiting case when R is much smaller than h. Minimizing E yields a height H (often referred to as Jurin's height) of the capillary rise:

$$H = \frac{2\gamma \cos \theta_E}{\rho g R} \tag{2.31}$$

where the impregnation parameter I has been replaced by its value given by Young's relation ($I = \gamma \cos \theta_E$).

Equation (2.31) warrants several comments. First, it agrees with Hauskbee's experimental observations (e.g., variation as $1/R$, lack of dependence on the outer pressure and on the thickness of the walls). Moreover, it describes capillary rise as well as descent (the sign of H is the same as that of $\cos\theta_E$). The height increases monotonically as the contact angle decreases, reaching a maximum H_M when $\theta_E = 0$ (for a 10-μm-wide tube filling up with water, H_M is about 1 m). One might even think that the rise could proceed beyond H_M, as might be reasoned in terms of energy, since nothing seems to preclude the parameter I defined in equation (2.29) from exceeding the value γ (this case simply corresponds to $S > 0$). In fact, the capillary rise does remain bounded by the value $H_M = H(\theta_E = 0)$ as written in equation (2.31). When $S > 0$, a microscopic film forms ahead of the meniscus. As far as the column is concerned, it is as though the parameter I were pegged at γ.

Finally, note that equation (2.30) was derived under the condition $R \ll H$, which implies that tube diameters are small compared to the capillary length ($R \ll \kappa^{-1}$). In practice, this means that diameters are in the submillimeter range. When this condition is not met, correction terms must be introduced in equation (2.31) [the first such correction term for $H(R)$ is of order R]. These correction terms were worked out in detailed mathematical developments first by Laplace, and subsequently by Poisson, Gauss, and Rayleigh in the 19th century.

2.4.3 Pressure Argument for the Capillary Rise

To find the height of the capillary rise, one can also use an argument in terms of pressures (Figure 2.18). In tubes that are very small compared with the capillary length, the leading meniscus (within the tube) is a portion of a sphere. The radius of curvature R of this sphere is equal to the radius R of the tube when the contact angle is equal to zero, and more generally to $R/\cos\theta_E$. The pressure immediately underneath the interface (point A in Figure 2.18) is given by Laplace's law (the curvature of a sphere of radius R is $2/R$):

$$P_A = P_0 - \frac{2\gamma\cos\theta_E}{R} \tag{2.32}$$

where P_0 designates the outer (atmospheric) pressure. The negative sign in equation (2.32) signifies that the interface is subject to an underpressure since its curvature is pointed toward the atmosphere (that is precisely what acts as a pump on the bath and causes the liquid to rise).

The pressure at point B (located at height $z = 0$) is P_0. The pressure difference between A and B is purely hydrostatic, point B bearing the weight of a liquid column of height H. Expressing the equilibrium of pressures gives

$$P_0 - \frac{2\gamma\cos\theta_E}{R} = P_0 - \rho g H. \tag{2.33}$$

The law of capillary rise (2.31) flows directly from equation (2.33).

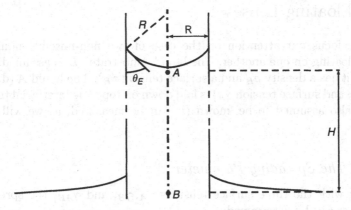

FIGURE 2.18. The hydrostatic pressure drop between points B and A offsets the Laplace underpressure at point A.

This type of argument is often the simplest approach for solving problems involving capillary rise. It should by now be obvious that capillarity will never be able to generate a jet (or a spring, contrary to what da Vinci believed). In order for a liquid to come gushing out of a tube, it would have to be in a state of overpressure, which would produce an inverted meniscus. But such a scenario would be incompatible with a rising liquid (which, as we know, implies an underpressure to balance the hydrostatic pressure). If one were to dip a tube of height h smaller than H in a liquid bath, the entire tube would fill with liquid and the meniscus would adopt whatever curvature offsets the hydrostatic pressure difference $-\rho g h$ with the bath.

What Bears the Weight of the Raised Liquid?

Finally, the fluid column resulting from capillary rise can be described in terms of the forces at equilibrium. There exists a force F that supports the weight of the column, which can be expressed as

$$F = \rho g \pi R^2 H. \tag{2.34}$$

Replacing H by its value derived in equation (2.31) yields

$$F = 2\pi R \gamma \cos \theta_E. \tag{2.35}$$

The length of the contact line is $2\pi R$. The capillary force f per unit length supporting the fluid column is given by

$$f = \gamma \cos \theta_E = \gamma_{SO} - \gamma_{SL} \equiv I. \tag{2.36}$$

In a total wetting regime $(I > \gamma)$, $F = 2\pi R \cdot (\gamma_{SO} - \gamma_{SL})$ exceeds the weight of the liquid column, which is always given by $2\pi R \gamma$. As a matter of fact, F also supports the weight of the wetting film to be discussed later on in chapter 4.

2.5 Floating Lenses

We now focus our attention on the case of two non-miscible liquids A and B floating on one another. The liquid "substrate" B is assumed to be denser (it has a density ρ_B and a surface tension γ_B). The liquid A (density $\rho_A < \rho_B$ and surface tension γ_A) is laid down on top. The interfacial tension γ_{AB} is also assumed to be known (it can be measured, as we will show shortly).

2.5.1 The Spreading Parameter

Starting with the three surface tensions γ_A, γ_B, and γ_{AB}, the spreading parameter can be determined:

$$S = \gamma_B - (\gamma_A + \gamma_{AB}).\tag{2.37}$$

If S is positive, wetting is total and the liquid spreads completely. Such is the case for a drop of PDMS deposited on the surface of water.

If S is negative, wetting is partial. A drop can form a lens or a large puddle, depending on the amount of liquid involved. That is precisely what one observes when depositing oil on water or molten glass on liquid tin.

2.5.2 The Shape of Floating Lenses $(S < 0)$

Floating drops distort the surface of a liquid substrate. To describe them accurately, it is necessary to include the distortions they cause around them. The competition between surface energy and gravitational energy now involves a total of three characteristic lengths (Figure 2.19b):

$$\kappa_A^{-1} = \sqrt{\frac{\gamma_A}{\rho_A g}}\tag{2.38}$$

$$\kappa_B^{-1} = \sqrt{\frac{\gamma_B}{\rho_B g}}\tag{2.39}$$

$$\kappa_{AB}^{-1} = \sqrt{\frac{\gamma_{AB}}{(\rho_A - \rho_B)g}}.\tag{2.40}$$

These lengths are associated with the $A/$air, $B/$air, and A/B interfaces, respectively.

Small Lenses $(R < \kappa_{AB}^{-1}, \kappa_A^{-1}$; Figure 2.19a)

For very small droplets, gravity is negligible and Laplace's internal pressure is constant. Surfaces have a uniform curvature. A droplet is made of two spherical caps that satisfy Neumann's vector relation all along the triple line:

$$\vec{\gamma}_A + \vec{\gamma}_B + \vec{\gamma}_{AB} = \vec{0}.\tag{2.41}$$

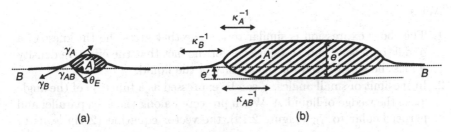

FIGURE 2.19. Floating lenses.

Large Lenses $(R > \kappa_{AB}^{-1}, \kappa_A^{-1};$ *Figure 2.19b)*

The drop is flattened by gravity. Its thickness e_c can be calculated either from the equilibrium of the forces involved or by minimizing the energy. The energy can be written in terms of the thickness e of the puddle and the immersed depth e' (Figure 2.19b):

$$F(e) = \left[\frac{1}{2}\rho_A g(e - e')^2 + \frac{1}{2}(\rho_B - \rho_A)ge'^2 - S\right]\frac{\Omega}{e} \qquad (2.42)$$

where Ω is the volume of the puddle and Ω/e is its surface area.

The sinking depth e' is such that Archimedes' force offsets the weight of the drop. It is possible to relate e and e' by writing that the hydrostatic pressures on either side of the liquid/liquid interface (assumed planar) are equal:

$$\rho_A ge = \rho_B ge'. \qquad (2.43)$$

If $\rho_B \approx \rho_A$, the lens is immersed much like an iceberg, whereas it protrudes if $\rho_A \ll \rho_B$.

Eliminating e' in equation (2.42) leads to

$$F(e) = \left[\frac{1}{2}\tilde{\rho}ge^2 - S\right]\frac{\Omega}{e} \qquad (2.44)$$

with $\tilde{\rho}$ being defined as

$$\tilde{\rho} = \frac{\rho_A}{\rho_B}\cdot(\rho_B - \rho_A). \qquad (2.45)$$

Upon minimizing the energy $F(e)$ at constant volume, one obtains

$$\frac{1}{2}\tilde{\rho}ge_c^2 = -S. \qquad (2.46)$$

This result was first derived by Langmuir.[9] It forms the basis of a very simple technique used to measure γ_{AB}, knowing γ_A and γ_B. We will see in chapter 7 (devoted to dewetting) that the thickness e_c also controls the stability of films of A floating on B.

Notes

1. The above expression is similar to the one that gives the thickness of a puddle on a solid substrate, except for the fact that the effective density $\tilde{\rho}$ must be substituted for the density of the liquid.

2. In the limit of small angles, S can be expressed as a function of the angle θ_E of the wedge of liquid A. When projected along the axes parallel and perpendicular to $\vec{\gamma}_B$ (Figure 2.19), the vector equation (2.41) leads to $\gamma_A \alpha_A = \gamma_B \alpha_B$, and $\gamma_B = \gamma_A(1 - \frac{\alpha_A^2}{2}) + \gamma_{AB}(1 - \frac{\alpha_B^2}{2})$. With $\theta_E = \alpha_A + \alpha_B$, the quantities α_A and α_B can be eliminated, yielding

$$\frac{1}{2}\tilde{\gamma}\theta_E^2 = -S \tag{2.47}$$

with $\tilde{\gamma}^{-1} = \gamma_A^{-1} + \gamma_{AB}^{-1}$. It follows that

$$e_c = \tilde{\kappa}^{-1}\theta_E \tag{2.48}$$

where $\tilde{\kappa}^{-1} = \sqrt{\frac{\tilde{\gamma}}{\tilde{\rho}g}}$. This analysis shows that for very flat floating lenses, the analogy with puddles on a solid substrate is complete if one replaces γ and ρ by $\tilde{\gamma}$ and $\tilde{\rho}$, respectively. For molten glass on liquid tin, $e_c = 6$ mm, which is comparable to the thickness of water floating on carbon tetrachloride.

Application: Measuring γ_{AB} with a Ruler. Measuring the radius of a floating lens gives its thickness if the volume is known. The parameter S can then be deduced from equation (2.46). Since in general γ and ρ are specified for a given liquid, γ_{AB} is readily deduced.

Example: A = PDMS and B = fluoroalkylsiloxane. For PDMS, $\rho_A = 0.97$ g/cm^3 and $\gamma_A = 21.0$ mN/m. For fluorinated oil, $\rho_B = 1.0$ g/cm^3 and $\gamma_B = 22.5$ mN/m. Assume we deposit drops of PDMS of increasing sizes. The volume can be determined by gravimetry to within one hundredth of a gram. For radii greater than a centimeter, the lens thickness becomes virtually independent of the radius, at which point $e_c = 1.32 \pm 0.05$ mm. This value enables us to calculate S with the help of equation (2.46). The result is $-S = 1.8 \pm 0.2$ mN/m, from which we deduce $\gamma_{AB} = 3.5 \pm 0.5$ mN/m.

2.6   Supplement on Techniques for Measuring Surface Tensions

There are several standard methods for measuring surface tensions when the interfaces involved are fluids (between a liquid and its vapor or between two liquids).

A surface tension is an energy per unit area that manifests itself by minimizing surface areas. A first class of measurement techniques consists in analyzing the shape of drops placed in specific situations and in deducing the value of the surface tension by adjusting the appropriate parameters in a mathematical model.

We also know that a pressure (the Laplace pressure) is associated with a curved surface, and that the pressure in question is proportional to the surface tension. Pressure measurements in the interior of drops will also provide information about the surface tension, provided that their curvature can be determined adequately.

As we have seen in chapter 1, an energy per unit area is equivalent to a force per unit length. Accordingly, a third natural class of techniques involves the measurement of the capillary force acting on a solid object placed at the surface of a liquid.

It is far more difficult to measure the surface tension of solids. In principle, one could envision tugging at a point of the surface, but the measured force would then be totally dominated by the elastic properties of the solids, without any ready means to isolate the contribution of the surface. Nevertheless, it is worth mentioning some recent work that did yield values of the surface tension at solid/vapor and solid/liquid interfaces in an ideal system.

Finally, the goal is oftentimes more to determine *changes* in surface tension (related, for instance, to temperature variations or to a gradual contamination of the surface by surfactants) than it is to determine absolute values. In most cases, the methods we describe can be used in a dynamical mode. As such they are eminently suitable for these types of measurements.[10] A good review on this topic can be found in the book by Adamson.[2]

2.6.1   The Shape of Drops

The shape adopted by a drop stems from a compromise between the effect of surface tension, which favors a sphere, and gravity (or any other force field), which will cause distortions. For any given situation, analyzing the shape of the drop ought to make it possible to extract the surface tension. Several methods currently used to measure surface (or interfacial) tensions are based on this principle, such as the sessile drop method or the pendant drop method. The latter is the most commonly used.[10–13]

2.6.1.1   The Pendant Drop Method

The basic principle of this method is to let a drop dangle loose at the end of a fine capillary tube. The drop takes on the shape of a light bulb (Figure 2.20).

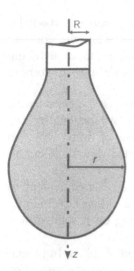

FIGURE 2.20. Pendant drop dangling at the end of a tube of inner radius R.

The tally of the pressure forces in equilibrium at every point of the drop includes a Laplace term and a hydrostatic term. If C is the curvature of the drop's surface, γ its surface tension, and ρ its density, the equilibrium condition reads

$$\gamma C = \rho g z. \tag{2.49}$$

The curvature can be expressed in a cylindrical coordinate system (the drop is symmetric around the z-axis). After defining the quantities $r_z = \frac{dr}{dz}$ and $r_{zz} = \frac{d^2 r}{dz^2}$, we have

$$C = -\frac{r_{zz}}{(1+r_z^2)^{3/2}} + \frac{1}{r(1+r_z^2)^{1/2}} \tag{2.50}$$

where the two principal radii of curvature have been written out explicitly at each point $r(z)$ on the surface of the drop.

Equations (2.49) and (2.50) can be solved numerically. The strategy is to treat the surface tension as an adjustable parameter and tweak its value until the results agree with experimental observations. The accuracy that can be achieved is satisfactory (about 1%). Furthermore, the principle discussed here can readily be extended to interfacial tension measurements on the condition that the density of the liquid be replaced by the difference between the densities of the two fluids in equation (2.49).

A pendant drop occasionally breaks loose. This occurs when its weight W exceeds the capillary force that holds it in place (Tate's law). The capillary force is at most equal to $2\pi R\gamma$, where R is the inner radius of the tube. In principle, one could determine the value of γ simply by weighing a drop that has fallen off. While the technique appears extremely straightforward at first blush, its practical implementation runs into a number of complications because of the dynamics of drop separation. Drops stretch, and necks form and get distended, so that in the end only a fraction αW of the original weight breaks off (α is typically about 60%), while the remainder remains stuck on the capillary. The factor α is a function of the ratio R/R_g (R_g denoting the radius of the falling drop), which can be found in tabular form in specialized texts. With these caveats in mind, it is indeed possible to deduce values of surface tensions from such measurements.

Despite the complications just discussed, experiments turn out to be remarkably reproducible for a given capillary tube and a given liquid. The technique provides a convenient means for producing well-calibrated drops, the radius of which is given by

$$R_g = \left(\frac{3}{2\alpha} \kappa^{-2} R \right)^{1/3} \tag{2.51}$$

where κ^{-1} is the capillary length. This equation can be understood by balancing the capillarity force $2\pi\gamma R$ against the droplet weight. By playing with the surface tension and/or the diameter of the tube, it is possible to adjust the radius of the drops to some degree. Yet, because of the $1/3$ power dependence, drops always end up having a radius in the millimeter range regardless of the specific values of R and κ^{-1}.

Finally, it is worth mentioning that the complex dynamical process of drop separation, particularly when the liquid is viscous, has been the object of several detailed studies in recent years.[14] These investigations have revealed the formation of secondary structures and were able to elucidate the shape the liquid filament takes when it pinches off just prior to separation. Figure 2.21, reproduced from Shi, Brenner, and Nagel, shows a drop whose viscosity is 10 times that of water, at the instant it separates from a

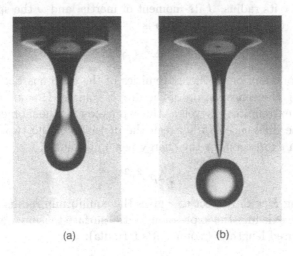

(a)   (b)

FIGURE 2.21. Drop of liquid as it detaches from a capillary tube. The shape it takes on is remarkable, particularly in the pinch-off area (courtesy S. Nagel and X. D. Shi).[15]

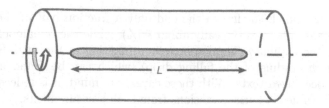

FIGURE 2.22. Spinning drop in a drum filled with a denser liquid. The rotation causes the drop to stretch lengthwise.

capillary 1.5 mm in diameter.[15] One can see quite clearly the appearance of a viscous filament, as well as its constriction where it is about to break off. The length of the filament as well as the secondary structures that form subsequently depend primarily on the viscosity of the liquid.

2.6.1.2 Spinning Drops

An interesting variation on the theme is the method of the spinning drop. The technique consists in placing a drop (or a bubble) in a cylindrical drum filled with another, denser liquid (Figure 2.22). The container is set in rotation at a few thousand revolutions per minute. Since the drop is not as dense as the surrounding liquid, it settles at the center and stretches along the cylinder's axis, adopting the shape of a filament, the length of which (typically a few centimeters) can easily be measured with good accuracy.

By ejecting the surrounding liquid toward the edges of the drum, the centrifugal force causes the drop to stretch lengthwise. At the same time, the surface tension γ between the two fluids opposes that trend. The energy of the stretched drop thus includes two competing terms. If L is the length of the drop, r its radius, J its moment of inertia, and ω the spinning rate of the drum, the energy of the drop is

$$E = \frac{1}{2}J\omega^2 + \gamma 2\pi r L \qquad (2.52)$$

where the contribution of the extremities of the drop has been neglected. After writing the conservation of the drop's volume ($\Omega = \pi r^2 L$, again neglecting the extremities) as well as the expression for the moment of inertia involving the difference $\Delta\rho$ between the densities of the two liquids, one arrives at an expression for the energy per unit volume:

$$\frac{E}{\Omega} = \frac{1}{4}\Delta\rho\omega^2 r^2 + \frac{2\gamma}{r}. \qquad (2.53)$$

Minimizing E with respect to r gives the equilibrium radius of the drop, from which we deduce the expression for the surface tension γ as a function of the measured length L (Vonnegut's formula):

$$\gamma = \frac{1}{4\pi^{3/2}}\Delta\rho\omega^2 \left(\frac{\Omega}{L}\right)^{3/2}. \qquad (2.54)$$

The resistance a drop (or a bubble) exhibits to stretching is striking when one inserts actual numbers into equation (2.54). Consider a drop of volume 1 mm^3, a difference in density of 100 kg/m^3, and a rotation rate of 10^3 rad/s. If one records a stretched length of 1 cm, one finds that the interfacial tension between the two fluids is only about 3 mN/m, a value that is typical of water/oil systems in the presence of surfactants. This particular approach is the method of choice for measuring very low interfacial tensions (of the order of a few mN/m or less). A great advantage of the method is that it is one of the few that does not involve contact with a solid, which would complicate the analysis and perturb experiments because of wetting problems.

2.6.2 Pressure Measurements

A widely used technique for measuring the surface tension is known as the method of maximal pressure of a bubble. Few people are aware that one of its pioneers was a young physicist named Erwin Schrödinger, back in 1915. The principle is quite simple. A capillary tube is partially immersed in a liquid bath down to a depth h much larger than its radius. Air is blown into the tube while the pressure is being measured when the bubble is forming.[10,13]

The pressure measured at the end of the tube is

$$P(R) = P_0 + \rho g h + \frac{2\gamma}{R}. \tag{2.55}$$

The radius of curvature R of the interface between the liquid and the vapor (spherical as long as $R \ll \kappa^{-2}/h$) varies as more air is being blown in. Figure 2.23 schematically depicts a few successive stages of the interface. At first, R decreases, then goes through a minimum (when $R = R$), before increasing again. It follows that the measured pressure itself goes through a

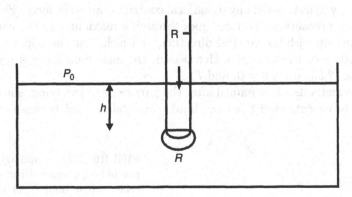

FIGURE 2.23. Measuring the maximal pressure of a bubble. When air is blown into a tube immersed into a liquid, the pressure at the end of the tube goes through a maximum when the bubble is hemispherical.

maximum equal to $P(\mathsf{R})$, from which the surface tension γ can be deduced. This method is precise and robust (it works even at very high temperatures, allowing experiments with metals and molten glass). In addition, it offers the opportunity to refresh the interface at will since the bubble eventually breaks off, dragging with it any potential contaminants that happen to get adsorbed on the surface, and a brand new interface is created.

2.6.3 Force Measurements

Finally, measuring the forces involved is another way to determine surface tensions. The procedure here consists of bringing into contact a liquid and an object with a well-defined geometry (often a plate or a fiber; Figure 2.24). The objective is to measure the capillary force exerted on the object (Wilhelmy's method). As we have seen in equation (2.35), this force can be written as

$$F = p\gamma \cos\theta \qquad (2.56)$$

where p designates the perimeter of the contact line and θ is the contact angle. The perimeter is easy to measure ahead of time, but one drawback of equation (2.56) is that it contains two unknowns, namely, γ and θ. Several tricks exist to eliminate θ. The first is to use a solid with a high surface energy, wettable by all usual liquids, in which case equation (2.56) reduces to $F = p\gamma$. Unfortunately, such a solid has a strong affinity toward contaminants since any speck of dust adsorbed on the solid lowers the surface energy. It becomes essential to clean the surface carefully before use. A popular choice of material is platinum, whose surface is easily regenerated by a flame.

The most common solution to the problem is to work in the retraction mode. The technique is based on removing the solid from the liquid bath very slowly so as to avoid any dynamical contribution to the forces involved. With this precaution, the force goes through a maximum as the capillary force lines up with the vertical direction, at which time the apparent contact angle goes through zero. Here again, the maximum force is given by equation (2.56) when $\theta = 0$, and $F_{\max} = p\gamma$.

These methods, often named after the particular object (ring, spur, etc.) that is being extracted from the liquid, are widely used because of their

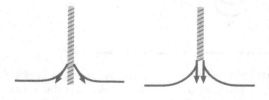

FIGURE 2.24. Capillary force exerted on a small object in contact with a liquid. As the object is extracted, the force goes through a maximum.

good accuracy, typically of the order of 1%. It is easy to control the temperature of the vessel containing the bath and to study phenomena as a function of temperature. As long as the relevant phenomena are not too fast, the technique also provides a means for studying the dynamical properties of adsorption at a liquid/air interface, making it possible to determine the corresponding surface tension.

2.6.4 Soft Solid Interfaces

All the measurement techniques reviewed so far have one common thread: They are all based on monitoring the distortions of interfaces. The key is to determine how surface tension resists such distortions. The fundamental reason that these methods are so simple to use is that up to this point we have been dealing exclusively with fluids. The interfaces considered so far have been between a liquid and a vapor or between two liquids. In such cases, surface tension is the dominant mechanism, even though gravity may at times complicate matters.

As such, these methods cannot be readily applied to hard interfaces, whether between a solid and a vapor or between a solid and a liquid, because the elastic energy stored in the solid far exceeds the interface energy associated with any distortion. As a result, measuring the surface tension of solids is generally perceived as an impossible task. Nevertheless, there exist a few examples of direct or indirect measurements that are worth mentioning.

The JKR Test

In the early 1990s, Chaudhury and Whitesides proposed using a standard adhesion test developed in 1971 by Johnson, Kendall, and Roberts (known as the JKR test) to determine interfacial tensions between a solid and a liquid or a solid and a vapor.[16] The procedure consists of bringing a hemispherically shaped solid, of radius typically 1 mm, in contact with a planar surface made of the same material (Figure 2.25).

Experimental observations show that contact occurs not at a single point but over a disk of radius l, because such a configuration lowers the overall energy (an extended contact reduces the free surface area of the solid). The first thing to measure is the radius, which is generally done with a microscope viewing the scene from below if the materials are transparent (as shown in Figure 9.27).

While the surface energy of the solid favors an extended contact, elasticity strenuously opposes it. The JKR formula expresses the fact that the radius l results from a compromise between these two forms of energy.[17] It reads

$$l = \left(\frac{6\pi W R^2}{K} \right)^{1/3} \tag{2.57}$$

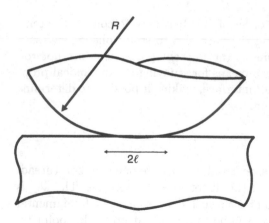

FIGURE 2.25. The JKR test. A sphere and a plane are brought into contact. The contact surface area is a function of the load on the sphere.

where W is the adhesion energy per unit area associated with the contact, R is the radius of the sphere, and K is the elastic modulus (an energy per unit volume) characteristic of the solid ($K = \frac{4}{3}\frac{E}{1-\nu^2}$, where E is Young's modulus and ν is Poisson's coefficient; in the case of an incompressible elastomer, we have $\nu \cong \frac{1}{2}$ and $K = \frac{16}{9}E$). For relatively soft solids (such as elastomers), for which this technique is ideally suited, K is small, of the order of 1 MPa (as compared with 10^5 MPa for glass) and the length l becomes readily measurable. For a silicone elastomer (such as bathtub caulk), l turns out to be about one hundred microns. Equation (2.57) can be generalized to the situation when a weight F is added to the hemispherical probe. In this case, the contact length l is an increasing function of F and the value of the modulus K can be extracted from the experimentally measured dependence. In the limit $F \gg WR$, the size l of the contact is given by Hertz's law ($l^3 \cong \frac{FR}{K}$).[18] The technique yields the value of K.

The next step is to measure the contact length l for various values of the sphere radius R. Figure 2.26 shows the results obtained by Chaudhury and Whitesides.[16] The data confirm that the contact radius does indeed increase as $R^{2/3}$. The slope of the graph determines the adhesion energy W per unit area, since K is known at this point. When the surrounding medium is air, the following simple relation holds:

$$W = 2\gamma_{SO} \tag{2.58}$$

where γ_{SO} is the surface tension of the solid. When applied to a silicone elastomer, this method gives $\gamma_{SO} = 21.2$ mN/m with an accuracy of about 4%.

As an extension of this technique, we can introduce into the system a vapor (from a substance other than the solid) prone to getting adsorbed on the surface of the solid, allowing us to measure the surface tension γ_{SV} of the solid/vapor interface.

Nothing prevents us from immersing the system in a liquid, allowing us this time to measure the surface tension γ_{SL} between the solid and the liquid. In chapter 9, we shall discuss the dynamics of how such contacts immersed in a liquid get established. Figure 2.27 shows on the abscissa values

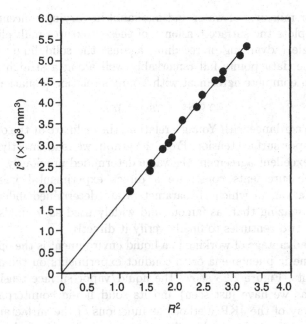

FIGURE 2.26. Cube of the contact radius l^3 (units are 10^{-3} mm^3) as a function of R^2, the square of the radius of the sphere (units are mm^2).[16]

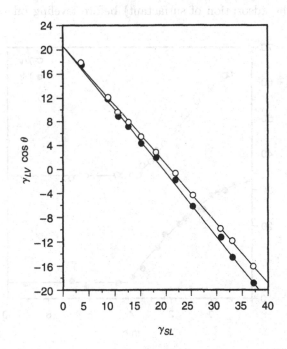

FIGURE 2.27. Surface tension of the liquid $\times$ contact angle (both advancing and receding, hence the two lines) plotted against the surface tension at the solid/liquid interface. Both quantities are in mN/m.[16]

obtained for mixtures of water and methanol in varying concentrations.[16] The graph plots the surface tension γ of these mixtures multiplied by the contact angle (advancing or receding) against the solid/liquid interfacial tension. The data points fall remarkably well along a straight line with slope -1, in complete agreement with Young's relation [equation (1.23)]:

$$\gamma \cos \theta = \gamma_{SO} - \gamma_{SL}. \tag{2.59}$$

Also in accordance with Young's relation, the ordinate at the origin gives the solid/vapor surface tension. From the graph, we read directly $\gamma_{SO} = 21$ mN/m, in excellent agreement the value determined previously.

These measurements constitute a direct experimental validation of Young's relation, in which all parameters are determined independently. It is truly amazing that, as famous and widely used as Young's law is, it took nearly two centuries to finally verify it directly.

Another advantage of working in a liquid environment is the opportunity to study kinetic phenomena or to conduct experiments on the adsorption of surfactants. Figure 2.28 shows the liquid/vapor surface tension (easily measured, as we have just seen) and its solid/liquid counterpart (determined by way of the JKR method) as functions of the surfactant concentration in a system comprised of silicone, water, and heptaethyleneglycoldodecylether.[19]

The graph shows that the surface tension decreases at both interfaces (because of the adsorption of surfactant) before leveling off at a concen-

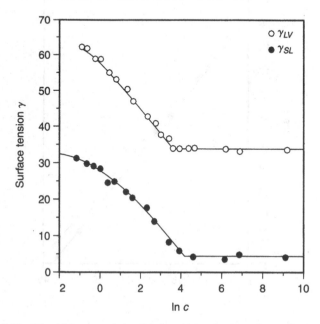

FIGURE 2.28. Liquid/vapor and solid/liquid surface tensions (both in mN/m) as functions of surfactant concentration.[19]

tration known as critical aggregation (or "micellar") concentration. These various behaviors will be studied in more detail in chapter 8.

References

[1] G. Debregeas and F. Brochard-Wyart, *J. Colloid Interface Sci.* **190**, 134 (1997).

[2] A. W. Adamson, *Physical Chemistry of Surfaces* (New York: John Wiley and Sons, 1976).

[3] C. Allain, D. Ausseré, and F. Rondelez, *J. Colloid Interface Sci.* **107**, 5 (1985).

[4] R. Fondecave, *Doctoral thesis, University of Paris* (1997); R. Fondecave and F. Brochard-Wyart, *Macromolecules* 31, 9305 (1998).

[5] C. Andrieu, D. Chatenay, and F. Rondelez, *C. R. Acad. Sci. (Paris)* **320**, 351 (1995).

[6] C. Redon, *Doctoral thesis, University of Paris* (1991).

[7] D. F. James, *J. Fluid Mech.* **63**, 657 (1974); L. L. Lo, *J. Fluid Mech.* **132**, 65 (1983).

[8] K. M. Takahashi, *J. Colloid Interface Sci.* **134**, 181 (1990).

[9] I. Langmuir, *J. Chem. Phys.* **1**, 756 (1933).

[10] E. I. Frances, O. A. Basaran, and C. H. Chang, *Current Opinion in Colloid and Interface Science* 1, 296 (1996).

[11] O. I. Del Rio and A. W. Neumann, *J. Colloid Interface Sci.* **196**, 136 (1997).

[12] J. P. Garandet, B. Vinet, and P. Gros, *J. Colloid Interface Sci.* **165**, 351 (1994).

[13] K. J. Mysels, *Colloid and Surfaces* **43**, 241 (1990).

[14] J. Eggers, *Rev. Mod. Phys.*, **69**, 865 (1997).

[15] X. D. Shi, M. P. Brenner, and S. R. Nagel, *Science* **265**, 219 (1994).

[16] M. K. Chaudhury and G. M. Whitesides, *Langmuir* **7**, 1013 (1991).

[17] K. L. Johnson, K. Kendall, and A. D. Roberts, *Proc. Roy. Soc. London* **A324**, 301 (1971).

[18] H. Hertz, *Miscellaneous Papers* (London: McMillan and Co, 1896).

[19] H. Haidara, M. K. Chaudhury, and M. J. Owen, *J. Phys. Chem.* **99**, 8681 (1995).

3

Hysteresis and Elasticity of Triple Lines

3.1 Description of Phenomena

3.1.1 Advancing and Receding Angle

When we place a liquid drop on a clean, planar, solid surface, we can observe a contact angle θ_E, which is precisely the angle contained in Young's formula. Quite often, though, the surface is marred by defects that are

- either chemical (stains, blotches, blemishes)
- or physical (surface irregularities).

On a non-ideal surface, the static contact angle turns out not to be unique. If, for instance, we inflate a drop (Figure 3.1a), the contact angle θ can exceed θ_E without the line of contact moving at all. Eventually, θ reaches a threshold value θ_A beyond which the line of contact finally does move. θ_A is referred to as the *advancing angle*.

Likewise, when deflating a drop (Figure 3.1b), θ can decrease down to a limiting value θ_R known as the *receding angle*. When $\theta = \theta_R$, the line of contact suddenly shifts. Generally speaking, it can be said that the observed angle θ depends on the way the system was "prepared." The contact angle *hysteresis* is defined as the difference between the limiting angles θ_A and θ_R.

Hysteresis is what makes it possible to capture a liquid column suspended in a vertical capillary of radius R (Figure 3.2). The upward pull exerted by the upper line of contact is $\gamma \cos \theta_1 2\pi R$. The downward force exerted by the second line of contact is $\gamma \cos \theta_2 2\pi R$. These two forces must exactly

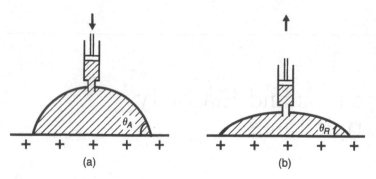

FIGURE 3.1. (a) Advancing angle when the drop is inflated; (b) receding angle when the drop is deflated.

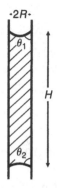

FIGURE 3.2. Liquid column trapped in a vertical capillary tube.

balance the weight of the column $\rho g \pi R^2 H$, from which we deduce

$$\frac{2\gamma}{R}(\cos\theta_1 - \cos\theta_2) = \rho g H. \tag{3.1}$$

For an equilibrium to exist, the upper line must be able to resist a receding movement, which implies $\theta_1 > \theta_R$ and $\cos\theta_1 < \cos\theta_R$. Similarly, the lower line must resist an advancing movement, which means $\theta_2 < \theta_A$ and $\cos\theta_2 > \cos\theta_A$. Consequently, equilibrium is possible only if

$$\frac{2\gamma}{R}(\cos\theta_R - \cos\theta_A) > \rho g H. \tag{3.2}$$

Without hysteresis ($\theta_R = \theta_A$), there can be no equilibrium (this is "Bertrand's theorem," cited by Bouasse).

As a rough guideline, on a "good" surface, the difference $\theta_A - \theta_R$ is small ($< 5°$). On a surface that is rough and/or dirty, that difference can exceed 50°.

Dettre and Johnson were the first to carry out a detailed analysis of hysteresis on surfaces with varying roughnesses.[1] They used paraffins and

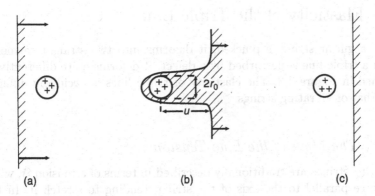

FIGURE 3.3. Mechanism of hysteresis (planar surface viewed from above). The shaded regions represent the liquid.

fluorinated chains. They started with compacted powders and decreased the surface roughness by means of successive heat treatments. They found that, as the roughness increases, the magnitude $\theta_A - \theta_R$ of the hysteresis increases at first and goes through a maximum before decreasing suddenly. An explanation for this behavior will be provided in chapter 9, which is devoted to the properties of "textured media."

3.1.2 Pinning of the Triple Line

The fundamental mechanism of hysteresis is illustrated in Figure 3.3. The figure depicts the retraction of a drop at a localized blemish that is more wettable than the rest of the surface. While retracting, the triple line moves toward the right and encounters the blemish (stage a). If the blemish is strong enough, it will "pin" the line locally and force it to stretch and warp (stage b). Nadkarni and Garoff have observed such distorted lines in detail.[2] Eventually, with enough force, the line breaks off the blemish (stage c). The break off dissipates energy even though on a macroscopic scale the line may move very slowly.*

If instead of imposing on the line a certain displacement velocity, we exert on it a macroscopic force per unit length given by

$$F = \gamma(\cos\theta - \cos\theta_E). \tag{3.3}$$

We find that break off is impossible to trigger below a threshold force. The line remains fixed for angles $\theta < \theta_E$.

We proceed next to discuss the energy associated with the distortion of the triple line.

---

*Note that the break off can be of one of two types depending on the final state. Either the zone near the defect remains dry, or it retains a captive wet area. Likewise, in the reverse process (advance), it is possible to have dry zones in the final state.

3.2   Elasticity of the Triple Line

When a guitar string is plucked, it deforms into two straight segments. When a triple line is perturbed by a defect, it deforms quite differently, as sketched in Figure 3.3. The elasticity of triple lines is decidedly different from that of vibrating strings.

3.2.1   The Myth of the Line Tension

Vibrating strings are traditionally described in terms of a tension $\Im$, which is a force parallel to the axis of the string, tending to stretch it. In this context, $\Im$ can often be thought of as an energy per unit length of the stressed string. It has been suggested that with small droplets (small but visible under a microscope), one can observe the effect of the tension $\Im$ associated with a triple line of contact. If true, Young's forces [equation (1.23)] are not enough to describe fully a drop of radius r deposited on a substrate. An additional force is called for. This force is to a curved line what Laplace's pressure [equation (1.5)] is to a curved surface. It is normal to the line and of magnitude $\Im/r$.* This gives rise to a correction term for the equilibrium angle given by

$$\theta - \theta_E \approx \frac{\Im}{\gamma r}. \qquad (3.4)$$

In most practical cases, the ratio $\Im/\gamma$ should be equal to the molecular distance a (a few Angströms), and the correction term $\Im/\gamma r$ would be of the order of 10^{-4} for a radius r of a few microns. Such sizes are at the limits of what can be observed by optical means.

As it turns out, several experimenters have reported abnormally large values of $\Im$ or of the length $\Im/\gamma$. Instead of the few Angströms one would normally expect, this length would be of the order of a micron! The result is, in our view, a spurious artifact traceable to the optical nature of the measurements. A possible source of error is illustrated in Figure 3.4. The flawed procedure predicts an *apparent* value of $\Im/\gamma$ comparable to the wavelength of the light, i.e., of the order of a micron. Large values of $\Im$ have no scientific reality; they amount to a myth.

We are about to see that in practice, the intrinsic tension of the line is obscured by a much more dominant phenomenon, namely, the *fringe elasticity* of a line of contact.

The line tension can be estimated from measurements other than optical ones, e.g., by studying the profile close to the contact line by atomic force

---

*This result can be justified by comparing the line energy of a circle of radius r to that of a circle of radius $r + dr$.

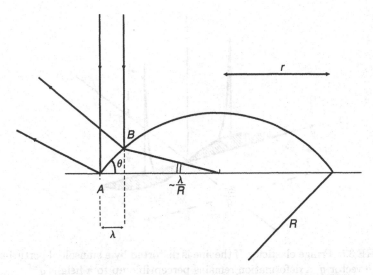

FIGURE 3.4. Optical reflection measurement of the contact angle. Because of the finite spatial resolution of a laser beam, the measurement is carried out at point B, rather than at the desired point A. The distance $\lambda = AB$ is of the order of an optical wavelength. This produces an error on θ of the order of $\lambda/R \sim \lambda r$. Should the error be interpreted as being due to a line tension $\Im$ (equation 3.4), the result is $\Im_{app} \approx \lambda\gamma$, giving results for $\Im_{app}$ that are roughly 1,000 times too high.

microscopy.[3] In this case, the deduced values ($\Im \approx 10^{-11}$ J/m) indeed lead to microscopic lengths for $\Im/\gamma$.

3.2.2   The Fringe Elasticity of the Line of Contact

The goal here is to calculate the distortion energy of a triple line of contact when the distortion has a small sinusoidal amplitude characterized by a wavevector q. For the sake of simplicity, we consider the particular case when the contact angle is 90° (Figure 3.5).

When at rest, the surface of the fluid is vertical and corresponds to the (y, z) plane. Consider now the case when the line is displaced by a small distance $u(y) = u_q \cdot \cos qy$ along the x-axis. The surface of the fluid is now distorted, with a local displacement $\zeta(y, z)$. At ground level, we have $\zeta(y, 0) \equiv u(y)$. Since we neglect gravity, the pressure inside the fluid is the atmospheric pressure. In accordance with Laplace's law, this implies a net total curvature of zero for the interface, which we write as

$$\frac{\partial^2 \zeta}{\partial y^2} + \frac{\partial^2 \zeta}{\partial z^2} = 0. \tag{3.5}$$

The proper solution to this equation is of the form

$$\zeta(y, z) = u_q \cdot \cos qy \cdot e^{-qz}. \tag{3.6}$$

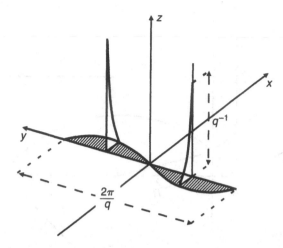

FIGURE 3.5. Fringe elasticity. If the line is distorted by a sinusoidal perturbation of wave vector q, a deformation remains perceptible up to a height q^{-1}.

Discussion of the Result

1. It should be emphasized that the local contact angle is no longer 90°. The situation no longer complies with Young's equilibrium. Indeed, at the substrate's level, we have

$$\left.\frac{\partial \zeta}{\partial z}\right|_{z=0} = -u_q q \cos qy \qquad (3.7)$$

$$\left.\frac{\partial \zeta}{\partial y}\right|_{z=0} = -u_q q \sin qy \qquad (3.8)$$

and the deviation from a 90°-angle is given by

$$|\nabla \zeta| = u_q q \qquad (3.9)$$

where ∇ is the gradient operator.

2. Equation (3.6) indicates that the distortion of a liquid's surface extends only over a characteristic thickness $1/q$; beyond that level, the surface reverts to a vertical plane.

The distortion energy per unit length along the y direction (in the approximation of a small $\nabla \zeta$) is

$$F = \int \frac{1}{2}\gamma \left\{ \left(\frac{\partial \zeta}{\partial z}\right)^2 + \left(\frac{\partial \zeta}{\partial y}\right)^2 \right\} dz = \frac{1}{4}\gamma q |u_q|^2. \qquad (3.10)$$

The crucial point to appreciate here is that the energy is proportional to q, rather than to q^2. Distortion energies $[\sim (\nabla \zeta)^2]$ do indeed vary as q^2, but they contribute to the integral only a "fringe height" (the height over which $\zeta \neq 0$), which is of order q^{-1}. This explains the linear dependence of the energy on q. It should be contrasted with the energy associated with the intrinsic line tension, which would read

$$\tilde{F} = \frac{1}{2}\Im \left\langle \left(\frac{du}{dy}\right)^2 \right\rangle = \frac{1}{4}\Im q^2 |u_q|^2. \tag{3.11}$$

The ratio of the two energies is

$$\frac{\tilde{F}}{F} \approx q\frac{\Im}{\gamma} \approx qa \tag{3.12}$$

where a is a molecular length. For practical wavelengths (optically observable), $qa < 10^{-3}$, and the conventional line tension is negligible.

The fringe elasticity we have just discussed has some remarkable consequences. In particular, it is responsible for the very special profile of a line that is slightly pinched. As indicated in Figure 3.3, let us assume that a weak force (due, for example, to a surface defect) is applied at point O of the substrate. What shape does the line of contact adopt? The answer requires solving an equation similar to (3.5), but one that includes a localized force f defined by

$$\gamma \left(\frac{\partial^2 \zeta}{\partial z^2} + \frac{\partial^2 \zeta}{\partial y^2}\right) = f \cdot \delta(x)\delta(y). \tag{3.13}$$

This is identical to the classical problem of the electrostatic potential created by a point charge in a two-dimensional space (y, z). If we define $r^2 = y^2 + z^2$, we have

$$\zeta = \frac{f}{\pi\gamma} \ln\left(\frac{r}{r_0}\right) \tag{3.14}$$

where r_0 is a cut-off radius comparable to the size of the defect. The reader may check equation (3.14) by writing the surface tension force $\gamma\frac{dz}{dr}$ integrated over a half circle of radius r.

The actual shape of the triple line of contact is obtained by simply letting $z = 0$ in the previous expression, which gives

$$u(y) = \zeta(y, 0) = \frac{f}{\pi\gamma} \ln\left(\frac{y}{r_0}\right). \tag{3.15}$$

This logarithmic shape is indeed what we can observe when we pull at a particular point of a line of contact (Figure 3.3). It is quite different from the shape predicted for a line in tension.

This entire discussion of the fringe elasticity has been greatly simplified by our choice of a particular contact angle ($\theta_E = \pi/2$). Nevertheless, the results we have just obtained qualitatively hold true for any value of θ_E. For a discussion of the general case, the reader is referred to the literature.[4]

3.3   Hysteresis Due to Strong, Sparse Defects

We start with the following assumptions:

- The defects on the substrates are far apart from one another.
- The defects are "strong," in the sense that each is capable of pinning a line (this criterion is discussed in more detail by Joanny et. al.[4]).

Under these conditions, it is straightforward to evaluate the threshold force per unit length required for the line to advance, which is given (at the macroscopic level) by the expression

$$F \equiv \gamma \cdot (\cos \theta_E - \cos \theta_A). \tag{3.16}$$

Assume now that we displace the line by a distance Δx. If n is the number of defects per unit area, the total number of defects that the line "snaps by" during its sweep is $n\Delta x$ (per unit length of the line). Each of these defects dissipates a "snapping" energy W, which is lost as viscous dissipation in the fluid. It follows that

$$F\Delta x = n\Delta x W. \tag{3.17}$$

All that remains to be done is to estimate the energy W. This can be done by noting that the elongation u (defined in Figure 3.3) of the line reaches a maximum value u_m.

We shall not presume here that the line distortions are small, contrary to what we did in our discussion of elasticity. As a matter of fact, when $\theta_E = \pi/2$, we can use the following direct line of reasoning. By pulling on a particular point of a fluid interface, we generate a profile with zero curvature as described in Figure 3.6. The profile has an analytical form in a semi-polar coordinate system (x, r) that is given by [see chapter 1, equation (1.18)]:

$$r = r_0 \cosh \left(\frac{x}{r_0} \right) \approx \frac{1}{2} r_0 \exp(x/r_0) \tag{3.18}$$

which holds when $x \gg r_0$.

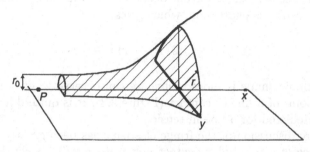

FIGURE 3.6. A vertical interface (yz plane) is deformed by pulling at the level of point P over a width $2r_0$ fixed by the size of the defect.

The minimum size r_0 will be comparable to the size of the defect. The "waist" radius r_0 can be related to the total force f exerted by the defect. At the waist radius, the surface tension γ pulls on a circle of perimeter $2\pi r_0$. It follows that the total force is $2\pi\gamma r_0$. Note, however, that we use only half the surface described by equation (3.18), namely, the half corresponding to $z > 0$. As a result, the force of the defect is only half that derived above, or $\pi\gamma r_0$.

The equation of the triple line of contact $u(y)$ is obtained by means of equation (3.18) after replacing x by u and r by y (since we then have $z = 0$):

$$u = r_0 \ln\left(\frac{2y}{r_0}\right) = \frac{f}{\pi\gamma} \ln\left(\frac{2y}{r_0}\right). \tag{3.19}$$

It is worth noting that this last equation agrees with equation (3.15), which was derived in the case of a small u. The total elongation u_m, defined in Figure 3.3, corresponds to $y \approx L/2$, where L is some average distance between adjacent anchor points along the line. We will not quantify here any more precisely the value of $L \propto 1/\sqrt{n}$, which in any event enters only through its logarithm:

$$u = \frac{f}{\pi\gamma} \ln\left(\frac{L}{r_0}\right) \approx K^{-1}f. \tag{3.20}$$

If we treat the logarithm as if it were a constant, we see that the line can be thought of as a spring of elongation u and rigidity K, with K given by

$$K \equiv \frac{\pi\gamma}{\ln(L/r_0)}. \tag{3.21}$$

The maximum elongation u_m corresponds to $f = f_m$, where f_m is the maximum force that the defect can exert just before the line snaps. We have $u_m = u(f_m) = K^{-1}f_m$. The energy stored in the fictitious spring under these conditions is

$$W = \frac{1}{2}Ku_m^2 = \frac{1}{2}\frac{f_m^2}{K} = \frac{f_m^2 \ln(L/r_0)}{2\pi\gamma}. \tag{3.22}$$

By substituting this expression back into equation (3.17), we get a formula giving the amplitude of the hysteresis:

$$\gamma(\cos\theta_E - \cos\theta_A) = \frac{nf_m^2 \ln(L/r_0)}{2\pi\gamma} \qquad (\theta_E = \pi/2). \tag{3.23}$$

For other values of θ_E, the formula is similar but more complicated (see ref. 4 for the case $\theta_E \ll 1$).[4] The salient conclusions are as follows:

- The amplitude of the hysteresis is proportional to the number of defects (in a diluted regime such that $L \gg r_0$).

- The amplitude varies as the square of the maximum pinning force f_m. This force f_m depends on the defect's wettability, size, and shape. It will be evaluated in the next section [equation (3.25)].

Avalanches. On occasions, the triple line jumps over distances greater than L (recall that L is an average distance between defects such that $nL^2 = 1$). The swept area A is roughly the product of an arc length y of the line by a perpendicular displacement x. Abnormal events such as we have just mentioned are characterized by $yx \gg L^2$. They have been observed mostly in a system composed of liquid helium 4/cesium/cesium vapor (cesium having been chosen because it gives partial wetting at low temperature). Prévost et al. sprinkled the surface with patches about 7 µm in size, which settled on the solid and played the role of strongly diluted defects.[5]

If the patches were uncorrelated, the probability of finding a swept area $A = xy$ would be given by a Poisson type of law $\exp(-nA) = \exp(-nxy)$, and the average value of x would be

$$x = \frac{1}{ny} = \frac{L^2}{y} \ll y \qquad \text{(for } y > L\text{)}. \tag{3.24}$$

In actuality, avalanches turn out to correspond to much larger values of x. This implies that the defects are in fact correlated. The occurrence of avalanches appears to always be associated with the existence of "empty spots" in the distribution of defects.

3.4   Surfaces With Dense Defects

3.4.1   A Realistic Example

Decker and Garoff have conducted a fairly systematic study of the shapes of lines of contact in the presence of numerous pinning sites.[6] The substrate they worked with was a fluoro-polymer, which they progressively degraded by exposing it to ultraviolet radiation. With water, the appearance of a typical line on a degraded substrate is illustrated in Figure 3.7.

Whereas isolated defects revealed by an atomic force microscope* are quite small (less than 1000 Angströms), the overall pattern suggests blemishes of dimension $\Delta y \sim 150$ µm. The depth of "valleys" between two blemishes is $\Delta x \sim 50$ µm. If we were to interpret each blemish as a macroscopic defect, we would expect (in light of the preceding section):

$$\Delta x \approx u_m \approx \frac{f_m}{\gamma} \tag{3.25}$$

---

*Atomic force microscopy (AFM) involves dragging an extremely fine probe tip along the surface to be characterized (see Figure 2.15 for the fabrication of the tip).

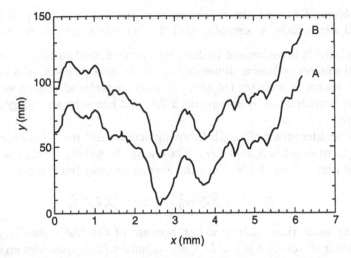

FIGURE 3.7. Relaxation of a line of triple contact pinned at defects (after Garoff and Decker).[6] (A) before displacement; (B) after a displacement of 500 μm.

or $f_m \approx \Delta\gamma_S \cdot \Delta y$, where $\Delta\gamma_S$ is the modulation of the surface energy of a blemish:

$$\Delta\gamma_S = (\gamma_{SL} - \gamma_{SO})_{blemish} - (\gamma_{SL} - \gamma_{SO})_{pure}. \tag{3.26}$$

This leads to

$$\Delta x \approx \frac{\Delta\gamma_S}{\gamma}\Delta y. \tag{3.27}$$

Inserting plausible numbers into this last equation suggests $\Delta\gamma_S \approx 1/3\gamma$.

The existence of "large blemishes" is hinted at by yet another observation contained in Figure 3.7. We get virtually the identical profile after displacing the wet region by a distance of 500 μm (for instance by tilting the substrate). This outcome would be impossible if we were dealing with small, uncorrelated defects. The size of the blemishes would have to be at least 500 μm.

Why did Garoff and Decker see large blemishes? It perhaps has something to do with the physical and/or chemical nature of the UV-induced degradation mechanism. For instance, the degraded regions may grow out of a few nucleation sites. The degraded areas can have a fine structure on a scale of 1,000 Å (the structure that is seen in AFM), but they may very well also have a spatial organization on larger scales due to the diffusion of a chemical species from the nucleation sites.

3.4.2 Small, Uncorrelated Defects

What happens if we are truly dealing with small, dense, uncorrelated defects, rather than with large blemishes? Theorists have pondered the question. Their current view is captured in the next problem.

Problem: Estimate the roughness of a triple line in the presence of small, dense defects, assuming that the line minimizes its energy.

Solution: It is convenient to describe the blemished surface in terms of small squares of lateral dimension r_0 to the side (the size of a defect). Each square is a defect (of area r_0^2) whose interfacial energy with the liquid is modulated by an amount $\pm\Delta\gamma$, and hence by an energy $\pm\varepsilon_1 = \pm\Delta\gamma r_0^2$.

Consider an arc of length L moving across a distance X. The number of squares swept is $N \approx XL/r_0^2$. The energy E_1 gained during the sweep is not proportional to N (since the average is zero) but, rather, to $\sqrt{N}$:

$$E_1 \approx -\sqrt{N}\Delta\gamma r_0^2 = -\Delta\gamma r_0(XL)^{1/2}. \tag{3.28}$$

At the same time, energy is lost because of the line's elasticity. The pertinent wavevector is $q \approx L^{-1}$, and equation (3.10) provides an elastic energy $\gamma q X^2$ per unit length. Over a length L, this translates to

$$E_2 \cong \gamma q X^2 L = \gamma X^2. \tag{3.29}$$

Minimizing $E_1 + E_2$ with respect to X gives

$$X^3 \cong \left(\frac{\Delta\gamma}{\gamma}\right)^2 Lr_0^2 \tag{3.30}$$

which shows that $X \propto L^{1/3}$. The exponent 1/3 has indeed been observed with liquid helium on a surface of cesium by Rolley et al.[7] Garoff et al., on the other hand, found a $X \propto L$ dependence, which better fits the description in terms of large blemishes, consistent with equation (3.27).

3.5 Two Cases Consistent With the Elasticity of Vibrating Strings

The complications due to the fringe elasticity disappear when the size (q^{-1}) of the fringes, is bounded by a physical cutoff.

3.5.1 Hele-Shaw Cells

The first example is the so-called *Hele-Shaw cells*, in which a fluid is confined between two parallel plates separated by a very small distance. The configuration is illustrated in Figure 3.8 [the plates are in (x, y) planes at $z = 0$ and $z = h$; the line is defined by the intersection of the vertical plane $(x = 0)$ with $z = 0$]. Again, we choose $\theta_E = \pi/2$ for the sake of simplicity.

We are looking for distortions of the line of contact that are slow on the scale of the thickness h of the cell. Under these conditions, it is legitimate

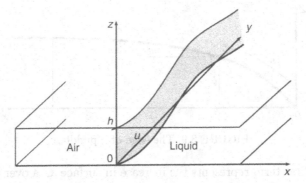

FIGURE 3.8. Horizontal Hele-Shaw cell. Before the deformation, an interface bounds a trapped drop between two horizontal plates. The deformation is the same at any height [$u(y)$ is independent of z].

to consider that the boundary of the liquid is vertical throughout and that it merely tracks the position of the line. This leads to an increase in energy given by

$$E = \frac{1}{2}\gamma h \left(\frac{du}{dy}\right)^2 = \frac{1}{2}\Im \cdot \left(\frac{du}{dy}\right)^2 \tag{3.31}$$

where $\frac{d}{dy} \ll \frac{1}{h}$ and $\Im = \gamma h$. It is evident on inspection of equation (3.31) that we have recovered the conventional elasticity of a guitar string. The behavior of such systems in the presence of localized defects has been studied both experimentally and theoretically.[8]

3.5.2 Puddle Edges

A second example concerns the boundary of puddles.* Near the contact line we have a wedge-shaped region extending only over a width of the order of κ^{-1} near the edges (see chapter 2). For wavelengths greater than κ^{-1}, one recovers the elasticity of a vibrating string. The details of the calculation follow.

The relevant geometry is described in Figure 3.9. The line is parallel to the y-axis. The length measured along the surface (curvilinear abscissa) is denoted s. The energy $\Im$ is the extra energy stored in the wedge. It includes both a capillary term and a gravitational term. The profile to be determined is denoted $z = \zeta(x)$. We have

$$\Im = \int dx \cdot \left\{\gamma \left(\frac{ds}{dx} - 1\right) + \frac{1}{2}\rho g \left(e - \zeta\right)^2\right\} \tag{3.32}$$

---

*Puddles have been defined in chapter 2. They are large drops deposited on a substrate and flattened by their own weight.

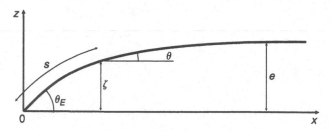

FIGURE 3.9. The edge of a puddle.

where the first term represents the increase in surface area over that of the same film kept flat, and the second term describes the energy required to remove the liquid near the puddle's edge, shape the film, and raise it to a height e.

To find the profile $\zeta(x)$ that minimizes $\Im$, we can directly write the equilibrium of the horizontal forces exerted on a slice of liquid (Figure 3.9):

$$\gamma(1 - \cos\theta) - \frac{1}{2}\rho g e^2 + \int_0^{\zeta(x)} p(x, z)\, dz = 0 \qquad (3.33)$$

where $p(x, z)$ is the hydrostatic pressure (excluding the atmospheric pressure p_0):

$$p(x, z) = \rho g(e - z). \qquad (3.34)$$

What is the justification for this particular pressure distribution? Note that the pressure is equal to p_0 in the flat portion of the puddle, immediately below the liquid surface ($z = e$), and increases by $\rho g e$ as the altitude dips below the surface down to the solid substrate ($z = 0$).

Equation (3.33) simply reduces to

$$2\sin\left(\frac{\theta(x)}{2}\right) = \kappa \cdot [e - \zeta(x)]. \qquad (3.35)$$

The integral giving the tension $\Im$ [equation (3.32)] then becomes

$$\Im = \int_0^\infty dx \cdot \gamma \left(\frac{1}{\cos\theta} - \cos\theta\right) = -\int_{\theta_E}^0 2\sin\left(\frac{\theta}{2}\right)\cos^2\left(\frac{\theta}{2}\right)\gamma\kappa^{-1}\, d\theta \qquad (3.36)$$

where we have introduced two successive changes of variable $dx = dz/\tan\theta$ and $dz = -d\theta \cdot \kappa^{-1}\cos\theta/2$. The procedure yields

$$\Im = \frac{4}{3}\gamma\kappa^{-1}\left(1 - \cos^3\frac{\theta_E}{2}\right). \qquad (3.37)$$

In the limit of small angles, equation (3.37) reduces to

$$\Im = \frac{1}{2}\gamma\kappa^{-1}\theta_E^2. \qquad (3.38)$$

Note the difference with eq (3.12). Here the line tension can be large and is observable.

3.5.3   Puddle Distortions

If one pulls on a particular point of a guitar string, it takes the shape of a triangle. Each part has zero curvature. The problem is not quite the same for a puddle because the puddle proper and the dry substrate are different media. Any macroscopic deformation of the puddle must keep the volume constant, which implies that the surface area of the wet region remains constant as well (since the thickness e of the puddle is volume-independent). This restriction is embodied in a Lagrange multiplier in the expression for the energy, which leads to a two-dimensional pressure p_2 (first introduced by Langmuir). The pressure is related to the edge curvature $1/R$ by analogy with Laplace's equation [see chapter 1, equation (1.5)]:

$$p_2 = \frac{\Im}{R} \tag{3.39}$$

where the curvature is counted positively if the puddle is a circle of radius R. A practical example is shown in Figure 3.10a, where one pulls on two opposite sides of a puddle with equal forces F. In such a case, the shape of the puddle consists of two half circles intersecting each other at an angle 2ϕ. Young's condition dictates that

$$F = 2\Im \cos \phi. \tag{3.40}$$

The size of the drop is such that its surface area remains equal to its initial value, which fixes the radius R. The puddle is the seat of a certain internal pressure $p_2 > 0$.

If, on the other hand, we start with a circular puddle and pull on four equidistant points, the outcome is a star-shaped structure (see Figure 3.10b) with four circular arcs, the curvature of which is inverted. In this case, the internal pressure p_2 is negative.

These types of experiments can all be conducted in an unglamorous kitchen, with nothing more than a puddle of oil sitting on the surface of water and a few bobby pins to pull on the edges.

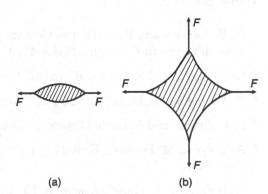

FIGURE 3.10. A puddle of oil deposited on water and stretched by either 2 pins (a) or 4 pins (b).

(a)   (b)

3.6   The Role of Thermal Fluctuations

In the preceding discussion, we have neglected the thermal fluctuations of the triple line. Such fluctuations are in fact unobservable on a macroscopic scale (roughly, for objects larger than micron). They can, nevertheless, play a role when one studies the hysteresis due to small defects. We proceed to discuss the possible role of thermal vibrations in overcoming an anchor point for the triple line.

Even when Young's force $\gamma(\cos\theta - \cos\theta_E)$ is quite weak, it is possible to see the line of contact move if the anchor points can be overcome by thermally activated jumps. The corresponding Boltzmann factor is $\exp(-W/kT)$, where kT is the thermal energy and W is the energy stored in the line at the instant of break off [equation (3.22)]. Based on equations (3.25) and (3.27), we have roughly

$$W \approx \frac{f_m^2}{\gamma} \approx \varepsilon^2 \gamma \Delta y^2 \qquad (3.41)$$

where $\varepsilon = \Delta\gamma_S/\gamma$ is a measure of the amplitude of the surface modulation and Δy is the size of a defect. An energy $W < \approx 20kT$ is required for jumps to be observed, which corresponds to

$$\Delta y < \left(\frac{20kT}{\gamma}\right)^{1/2} \varepsilon^{-1}. \qquad (3.42)$$

With $\gamma = 20$ mN/m and $\varepsilon = 0.1$, the requirement comes out to $\Delta y < 20$ nm.

Conclusion: Only for very small defects can thermal vibrations provide enough energy to overcome hysteresis. The reader is encouraged to consult the literature for a documented case involving very small defects (helium on cesium).[9]

References

[1] R. E. Johnson and R. H. Dettre, Contact Angle, Wettability and Adhesion, *Advances in Chemistry Series* **43**, 112 and 136 (1964).

[2] G. Nadkarni and S. Garoff, *Europhys. Lett.* **20**, 523 (1992).

[3] T. Pompe and S. Herminghaus, *Phys. Rev. Lett.* **85**, 1930 (2000).

[4] J. F. Joanny and P. G. de Gennes, *J. Chem. Phys.* **81**, 552 (1984).

[5] A. Prévost, M. Poujade, E. Rolley, and C. Guthmann, *Physica* B **280**, 80 (2000).

[6] E. Decker and S. Garoff, *Langmuir* **13**, 6321 (1997).

[7] E. Rolley, C. Guthmann, R. Gombrovicz, and V. Repain, *Phys. Rev. Lett.* **80**, 2865 (1998).

[8] M. Fermigier, J. F. Duprat, F. Goulaouic, and P. Jenffer, *C. R. Acad. Sci. (Paris)* **314**, 879 (1992).

[9] A. Prévost, E. Rolley, and C. Guthmann, *Phys. Rev. Lett.* **83**, 348 (1999).

4
Wetting and Long-Range Forces

4.1 Energy and Properties of Films

4.1.1 Transition From Macroscopic to Microscopic

Throughout the first three chapters, we have relied entirely on macroscopic concepts such as pressure, surface tension, and the like. A description based on such concepts is adequate as long as the dimensions of the drops and puddles under consideration exceed the range of interactions between molecules. If, on the other hand, we wish to study thin films (liquid pellicles of thickness e much less than a micron), we must take into account the range of interactions between molecules.

Let us place a film of thickness e on a solid substrate. If the film is thick (say, more than 100 nm), we can ascribe to it an energy $\gamma_{SL} + \gamma$ corresponding to independent contributions from two interfaces, namely, solid/liquid (γ_{SL}), and liquid/air (γ).

At the other extreme, when the film becomes very thin ($e \to 0$), we must recover the energy of the bare solid (γ_{SO}). The question we now turn our attention to is: What is the dependence of the energy per unit surface area between these two limits?

We begin by defining

$$Energy/m^2 = \gamma_{SL} + \gamma + P(e) \tag{4.1}$$

where $P(\infty) = 0$ and $P(0) = S = \gamma_{SO} - \gamma_{SL} - \gamma$. One might think that the function $P(e)$ should vary over a narrow range of thicknesses $e \approx a$, where a is the size of a liquid molecule. Actually, there often are long-range contributions to $P(e)$. In particular,

- When the various solid and liquid molecules interact via van der Waals forces (which vary as $1/r^6$), $P(e)$ turns out to vary much more slowly (as $1/e^2$).
- When we deal with solutions of polymers, the scale over which $P(e)$ varies can become as large as the size of a polymer coil.
- When the liquid is pure water, electrically charged layers form spontaneously at both interfaces. Such charged layers are subject to electrostatic interactions with a typical range of 100 Å.

All these effects will be discussed in more detail in section 4.2. For the time being, we will simply review a few general properties of such films, as they relate to $P(e)$.

4.1.2   Thickness Change and Disjoining Pressure

If we increase the thickness e of a liquid film by an amount de by adding a number of molecules dN (by unit surface area), we have $dN = de/V_0$, where V_0 represents the volume per molecule in the liquid phase. The energy varies by $\mu\,dN$, where μ is the chemical potential of the liquid. This potential includes a dominant term μ_0 due to volumetric effects, as well as a correction term $P(e)$ associated with the interfacial energy:

$$\mu\,dN = \mu_0\,dN + \frac{dP}{de}\,de. \tag{4.2}$$

Following Derjaguin's notation, we will define the *disjoining pressure* $\Pi(e)$ as the quantity

$$\Pi(e) \equiv -\frac{dP}{de} \tag{4.3}$$

When $P(e)$ is a decreasing function of e (favoring a thick film), $\Pi(e)$ is positive.

We can readily see that this pressure is related to the chemical potential via the relation

$$\mu = \mu_0 - v_0\Pi(e). \tag{4.4}$$

How can one measure the disjoining pressure? For usual, non-volatile liquids, the most common setup is that devised by Sheludko.[1] It is described in Figure 4.1.

The film under study is in equilibrium with a porous ring filled with the same liquid. The ring can be subjected to a hydrostatic pressure $p_H = \rho g H$,

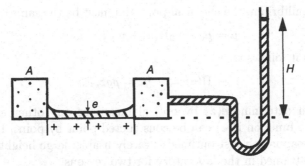

FIGURE 4.1. Measuring the disjoining pressure in the Sheludko method, using a porous ring A under a pressure adjustable by setting H.

where ρ is the density of the liquid. Equality of the chemical potentials in the film and in the ring requires that

$$\mu_0 - v_0\Pi(e) = \mu_0 + v_0 p_H \qquad (4.5)$$

from which we deduce $\Pi(e) = -p_H$. By measuring (for instance, via an optical technique) the thickness e as a function of p_H, one can construct the curve $\Pi(e)$.

It is also possible in principle to determine the disjoining pressure by monitoring the rise of a liquid film up a vertical wall (Figure 4.2).

We assume here that the retaining wall is totally wettable by the liquid. If a long-range attraction between the wall and the liquid exists, the molecules of liquid will tend to climb up against it so as to derive the most benefit. Naturally, the rise will be balanced out by the weight of the film. The starting point of our discussion is equation (4.4). Now the chemical potential is augmented by a mgz term, where m is the mass of a molecule, g is the acceleration of gravity, and z is the height above the surface of the

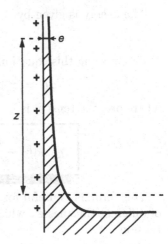

FIGURE 4.2. Measuring the disjoining pressure: Ascent of a film rising up a solid vertical wall.

bath. At equilibrium, the chemical potential must be the same everywhere:

$$\mu = \mu_0 - v_0 \Pi(e) + mgz = \mu_0 \qquad (4.6)$$

from which it follows that

$$\Pi(e) = \frac{m}{v_0} gz = \rho gz. \qquad (4.7)$$

If we can determine $e(z)$ by optical means or any other appropriate method, the function $\Pi(e)$ can be constructed point by point. In practice, though, the approach just outlined is rarely usable: large heights z cannot be readily attained in the laboratory for two reasons:

- A well-controlled surface is required over the entire height,
- the time needed to reach equilibrium can be prohibitive because the film is thin (typically 100 Å) and viscous friction is strong.

If a height $z = 1$ meter is achievable with a density $\rho = 1$ g/cm^3, we expect a disjoining pressure of about 0.1 atm. As happens quite often, the pressures of interest can be significantly larger (corresponding to thinner films), in which case Sheludko's method is the better approach.

4.1.3   Overall Stress in a Film

A macroscopic liquid film between a solid and air is subject to the sum of the interfacial tensions $\gamma + \gamma_{SL}$. What is the situation in the case of a thin film in which the corrective term $P(e)$ plays a role? To determine the corresponding tension $\gamma_{film}(e)$, we will consider the change in energy when the surface area of the film varies by an amount dA. But contrary to what we did in the previous case, we will this time implement the transformation keeping the number of molecules constant, that is to say, at constant volume Ae:

$$\frac{dA}{A} + \frac{de}{e} = 0. \qquad (4.8)$$

The energy is given by

$$E = A[\gamma + \gamma_{SL} + P(e)]. \qquad (4.9)$$

Differentiating this equation gives the energy variation

$$dE = [\gamma + \gamma_{SL} + P(e)]\, dA - A\Pi(e)\, de. \qquad (4.10)$$

Therefore, the tension is

$$\gamma_{film}(e) = \frac{dE}{dA} = \gamma + \gamma_{SL} + P(e) + e\Pi(e) \qquad (4.11)$$

This formula is useful for discussing the equilibrium state between films of different thicknesses, which must all experience simultaneously the same tension.

Note:
The above formula can also be derived from a different argument involving the internal pressure distribution $p(z)$ within the film (z being the distance to the solid substrate). At equilibrium, the pressure $p(z)$ is given by

$$p(z) = p_{atm} - \Pi(e) + \Pi(z). \qquad (4.12)$$

As an exercise, our reader may justify this formula by verifying that the force exerted by the solid on an element of volume at height z is $+d\Pi/dz$. Equilibrium implies that this force and the hydrostatic force $-\partial p/\partial z$ add up to zero.

By integrating $p(z)$ over the thickness of the film, one rediscovers the corrective term $P(e) + e\Pi(e)$ appearing on the right-hand side of equation (4.11).

4.1.4 Three Types of Wetting

4.1.4.1 Stability Condition

A few possible shapes for the dependence of the energy $P(e)$ are sketched in Figure 4.3. We should emphasize right away that a film can be stable only on condition that $P(e)$ has a positive curvature $[\ddot{P}(e) > 0]$. If the curvature is negative, a film of thickness e can split up into two films of thickness e_1 and e_2, respectively, occupying fractions α_1 and α_2 of the surface ($\alpha_1 + \alpha_2 = 1$). The energy then becomes $\alpha_1 P(e_1) + \alpha_2 P(e_2)$, which is lower than the initial $P(e)$, as indicated in Figure 4.3a. As a result, the film will lower its energy by splitting up into two parts. What is the final equilibrium state? The answer is provided by Figure 4.3c, which depicts the construction of a common tangent line. The two contact points correspond to two different thicknesses e' and e''.

Stated in words, films of thicknesses e' and e'' can coexist in the same sample. The fact that the tangent is common can be written as

$$\Pi(e') = \Pi(e'') = \Pi. \qquad (4.13)$$

This equation expresses the fact that the chemical potentials are equal. The boundary condition reads

$$e'\Pi(e') + P(e') = e''\Pi(e'') + P(e''). \qquad (4.14)$$

Given equation (4.11), this last formula states that the tensions are equal in both films.

Another important case is when the dry solid ($e = 0$) coexists with a liquid film (of thickness e_c). Here the stability of films is defined by a "simple tangent" connecting the dry point to the curve $P(e)$. An example appears in Figure 4.4a.

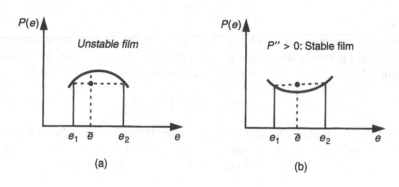

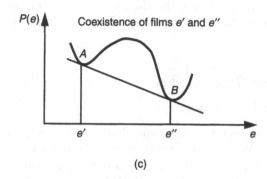

FIGURE 4.3. Stability criterion (a) $P'' < 0$; (b) $P'' > 0$; (c) coexistence of films e' and e''. Constructing the common tangent accounts for the possible coexistence of two films of thicknesses e' and e''.

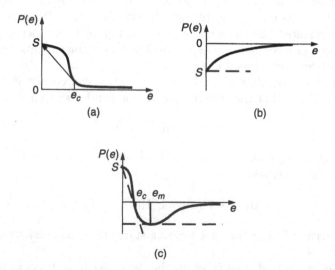

FIGURE 4.4. Different possible shapes of the potential $P(e)$: (a) $S > 0$ and P decreasing at long range (total wetting); (b) $S < 0$ and P increasing at long range (partial wetting); (c) $S > 0$ and P increasing at long range (pseudo-partial wetting).

4.1.4.2   Total Wetting

We are now in a situation where the spreading parameter S is positive and the functional dependence of the energy $P(e)$ has the shape illustrated in Figure 4.4a. The construction of a simple tangent line reveals a threshold thickness e_c. Films with a thickness greater than e_c are stable. Films with a thickness smaller than e_c are either metastable or downright unstable.

It is possible to have coexistence of a dry solid ($e = 0$) and a film of thickness e_c. We will call this particular type of film a "pancake." The quantity e_c is the thickness of such a pancake. Its value is obtained by constructing the simple tangent:

$$P(0) = S = P(e_c) + e_c \Pi(e_c). \tag{4.15}$$

This last equation can be interpreted as the equilibrium condition for the forces appearing in (4.11). The tension of the dry solid is γ_{SO}. The global tension of the pancake is $\gamma_{SL} + \gamma + P(e_c) + e_c \Pi(e_c)$. Equating the two leads directly to (4.15).

The thickness e_c is often quite small (a few Å), but it is very important from a conceptual point of view. When a drop spreads in a total wetting regime and the diagram of $P(e)$ looks like the graph in Figure 4.4a, the drop will proceed to thin down until its thickness reaches the pancake value e_c.

The only case where the above (macroscopic) discussion is significant is that in which the spreading parameter S is small. Since both $P(e)$ and $e\Pi(e) \to 0$ when $e \to \infty$, a small value of S implies a relatively thick film (a few tens of Å) and the macroscopic approach is valid. For larger S values (small e) a detailed molecular description of the film would be required.

4.1.4.3   Partial Wetting

Here, we have $S < 0$ and we deal with a dependence of the energy $P(e)$ of the kind depicted in Figure 4.4b. It remains possible to draw a "simple tangent" between two points on the curve, namely, the point at $[e = 0, P = S]$ (where the slope is undefined) and the point at infinity $[x \to \infty, P(e) \to 0]$. The physical meaning is that a dry solid ($e = 0$) can coexist with a drop ($e \to \infty$). This situation corresponds precisely to what we have previously called partial wetting, characterized by a contact angle θ_E such that

$$\gamma(1 - \cos\theta_E) = -S > 0. \tag{4.16}$$

4.1.4.4   Pseudo-Partial Wetting

Finally, we can have a curve of the type sketched in Figure 4.4c, characterized by a minimum of $P(e)$ for a finite thickness e_m. It is then possible to have a common horizontal tangent to the curve at point M and at a point at infinity, implying the coexistence of a film of thickness e_m and a macroscopic drop $(e \to \infty)$.*

What is the contact angle $\tilde{\theta}_E$ of the drop with the substrate? Based on equation (4.11), the equilibrium condition of the horizontal forces is $\gamma + \gamma_{SL} + P(e_m) + e_m \Pi(e_m) = \gamma_{SL} + \gamma \cos \tilde{\theta}_E$. Since $\Pi(e_m) = 0$, the equation just derived reduces to

$$\gamma(1 - \cos \tilde{\theta}_E) = -P(e_m). \qquad (4.17)$$

The equilibrium between a drop and a film prevails as long as there exists an ample supply of liquid. Should the available supply be insufficient to cover the entire substrate with a film of thickness e_m, then a drop placed down will spread completely (hence the name pseudo-partial wetting).

It almost goes without saying that the examples shown in Figure 4.4 hardly exhaust all possibilities for the dependence of $P(e)$ on e. We will later on encounter other behaviors in the context of "stratified" films.

4.2   The Nature of Long-Range Forces

4.2.1   van der Waals Forces

In vacuum, two molecules A and B separated by a distance r (greater than the range of chemical bonds) have an energy of attraction that varies as r^{-6}:

$$V_{AB} = -k\alpha_A \alpha_B \frac{1}{r^6} \qquad (4.18)$$

where α_A and α_B are the polarizabilities of molecules A and B, respectively, and k is a constant that depends very little on the nature of A and B. Equation (4.18) presupposes that the distance r is less than some upper limit λ of the order of 1000 Å.

From this formula, one can calculate the van der Waals energy of a film by adding up the liquid/liquid, liquid/solid, and solid/solid interactions.† One obtains in the macroscopic limit $(e > a)$:

$$P(e) = \frac{A}{12\pi e^2} \qquad (4.19)$$

---

*To understand this construction better, it may be useful to add to $P(e)$ the (very small) term due to gravity. This term is proportional to e^2 [equation (2.11)]. In such a case, there is clearly a common tangent between a point where e is very nearly e_m and a distant point where e is equal to the thickness of the puddle. Such a construction is discussed in detail in chapter 7 (Figure 7.3).

†This calculation is approximate because the interactions between dipoles are slightly modified by the dielectric properties of the medium. For a more rigorous derivation, see Israelashvili.[3]

where A has the dimension of an energy and is known as the *Hamaker constant*. Its numerical value is given by

$$A = \pi^2 k \bar{\alpha}_L (\bar{\alpha}_S - \bar{\alpha}_L). \tag{4.20}$$

In this last equation, the quantities $\bar{\alpha}_L$ and $\bar{\alpha}_S$ are the polarizabilities per unit volume of the liquid and the solid, respectively. The fact that A vanishes when $\bar{\alpha}_S = \bar{\alpha}_L$ should come as no surprise. Indeed, if the liquid and solid are identical, there no longer is any special energy associated with a particular thickness e.

From an order of magnitude standpoint, A is typically in the range of 10^{-20} to 10^{-19} J (comparable to the thermal energy equal to 4×10^{-21} J at 25°C). The sign of A, however, is of paramount importance. The most frequent case is when the solid has a higher polarizability than the liquid ($\bar{\alpha}_S > \bar{\alpha}_L$). This situation is said to involve a "high-energy" surface. The opposite case ($\bar{\alpha}_S < \bar{\alpha}_L$) corresponds to a "low-energy" surface. The latter situation most often applies to solids made of fluorocarbons.

Problem: What is the shape of a liquid film ascending a plate when van der Waals forces come into play?

Answer: The starting point is equation (4.4), which gives the chemical potential. Calculating $\Pi(e)$ with the help of equation (4.19), we find

$$\Pi(e) = +\frac{A}{6\pi e^3}. \tag{4.21}$$

Equating this to $\rho g z$, we deduce

$$e(z) = \left(\frac{A}{6\pi \rho g}\right)^{1/3} z^{-1/3}. \tag{4.22}$$

It often proves convenient to rewrite such a formula in a more explicit way. We do this by introducing a molecular length a via the relation

$$\frac{A}{6\pi \gamma} = a^2. \tag{4.23}$$

Typically, a is of the order of 1 Å. We can also reintroduce the capillary length κ^{-1} defined in chapter 2 ($\kappa^2 = \rho g / \gamma$). We then have

$$e^3(z) = \frac{a^2}{\kappa^2 z}. \tag{4.24}$$

For $z = 1$ m, $a = 1$ Å, and $\kappa^{-1} = 1$ mm, we find $e \sim 22$Å.

Supplementary problem: How high will the film rise?

Answer: There is a terminal zone where z suddenly drops to zero. In this zone, the film is no longer flat. Indeed, the curvature of the interface modifies the chemical potential. As a matter of fact, we already know the answer in light of our discussion on the equilibrium of a system composed of a film and a dry solid. We have seen that the film thickness must be the pancake thickness e_c. Inserting $e = e_c$ into equation (4.23) allows us to find the maximum height z_{max}:

$$\Pi(e_c) = +\rho g z_{max}. \tag{4.25}$$

With $e_c = 10$ Å, the corresponding value of $\Pi(e_c)$ is 1 atm, which yields $z_{max} \approx 10$ m.

Problem: A drop is placed on a fiber that has a small radius b. The fiber material is totally wettable by the liquid. What kind of wetting film will form under these conditions?

Answer: If the film has a thickness e_m, the radius of curvature of the coated surface is $b + e_m$. The chemical potential $\mu - \mu_0$ within the film includes a contribution μ_1 due to the curvature and another μ_2 due to the van der Waals forces. When calculating the change in energy as the thickness increases from e to $e + de$, one finds $\mu_1 = \frac{\gamma}{b+e} v_0$ and, by virtue of equation (4.4), $\mu_2 = -\Pi(e)v_0$, where as before v_0 is the volume of a molecule of liquid. Equating μ to the chemical potential μ_0 in the drop leads to $e_m \approx a^{2/3}b^{1/3} \ll b$. With $a = 1$ Å and $b = 10$ μm, the answer is $e_m \approx 50$ Å.

4.2.2 Case of Temperature-Dependent van der Waals Forces

In most situations, the Hamaker constant A is virtually independent of temperature. There are some special cases, however, when A does become temperature-dependent.

- Upon vigorous heating, the liquid coexists with its vapor (which has a polarizability $\bar{\alpha}_V$). Equation (4.20) no longer holds in this case and should be replaced by

$$A = \pi^2 k(\bar{\alpha}_L - \bar{\alpha}_V)(\bar{\alpha}_S - \bar{\alpha}_L). \tag{4.26}$$

As one approaches the critical point for the liquid/vapor pair, the difference $\bar{\alpha}_L - \bar{\alpha}_V$ decreases markedly and A becomes small.
- With a low-energy surface ($\bar{\alpha}_S$ small), it is possible to have $\bar{\alpha}_L > \bar{\alpha}_S$ at low temperature. Upon heating, however, the density of the liquid decreases, which means that $\bar{\alpha}_L$ decreases as well. It then becomes possible to end up with $\bar{\alpha}_L < \bar{\alpha}_S$. In other words, the constant A changes sign

at a certain temperature (see the published literature for a discussion of this transition).[2]

4.2.3   Van der Waals Interactions in Layered Solids: Surface Treatments

We now study in greater detail the surface treatments discussed in section 1.2.3.2. The standard way to apply a surface treatment is to deposit an antagonistic layer serving as an intermediary between the solid and the liquid so as to alter and possibly even reverse the wettability characteristics of the solid.

- The classic example of water on "floated" glass (which is exceedingly smooth) is a case where $\bar{\alpha}_S > \bar{\alpha}_L$. The parameters A and S are both positive and water spreads fully. To prevent water from spreading, the traditional treatment involves a layer of polarizability $\bar{\alpha}_1 < \bar{\alpha}_L$. A good candidate is a molecular coating deposited by silanization.
- Consider now the example of water (of polarizability $\bar{\alpha}_L$) on plastic (of polarizability $\bar{\alpha}_S = \bar{\alpha}_1 < \bar{\alpha}_L$). Equation (4.26) tells us that $A < 0$. Furthermore, $S < 0$ as well. We are therefore in a partial wetting regime. To render the surface wettable, one possible approach is to deposit a thin film of gold, which has a very high polarizability. What thickness d of the gold film is needed to completely conceal the original solid surface? Can water deposited on gold-coated plastic be tricked into "thinking" that it is resting on massive gold?

The qualitative answer is simple. Roughly speaking, the energy $P(e)$ results from sampling the solid over a thickness comparable to e. As long as $e \ll d$, the energy $P(e)$ is dominated by the thin solid coating (Hamaker constant A_1). If, on the other hand, $e \gg d$, the energy $P(e)$ samples the deepest part of the solid, i.e., the substrate itself ($A = A_2$).

Figure 4.5 sketches how the energy $P(e)$ evolves when glass is silanized. For bare glass, we have $S > 0$ and $A > 0$, which corresponds to total wetting. For silanized glass, $A_{eff} > 0$ for thick films. However, A_{eff} changes sign for thin films, which leads to $S < 0$ and puts us in a partial wetting regime.

Likewise, Figure 4.6 shows the evolution of $P(e)$ when a plastic surface is made wettable by coating it with a layer of gold. Prior to the surface treatment, we have, which implies $A_{eff} < 0$ and $S < 0$ (partial wetting). Following the surface treatment, A_{eff} is negative for thick films, and turns positive for thin films since $\alpha_1 > \alpha_L$. The $P(e)$ curve goes through a minimum at $e = e_m$, the value of which will depend on the thickness of the gold coating. One may immediately conclude that wetting will be of the pseudo-partial type. Since water has an affinity toward gold, it will spread as a thin film, but not as a thick one because it still senses the plastic

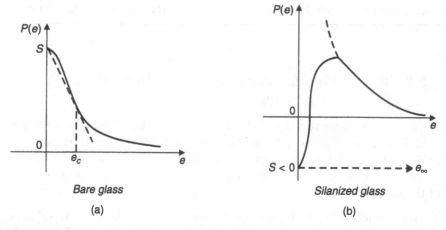

FIGURE 4.5. Energy $P(e)$ for (a) bare glass (total wetting) and (b) silanized glass (partial wetting).

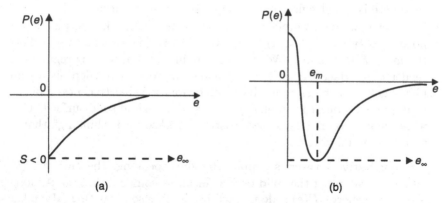

FIGURE 4.6. Energy $P(e)$ for (a) bare plastic (partial wetting) and (b) plastic coated with a film of gold (pseudo-partial wetting).

underneath the gold. At equilibrium, a drop of water will sit on gold-coated plastic, which is itself wetted by a very thin film.

In summary, a surface treatment relies on a thin coating of a material with a polarizability α_1. To render the liquid wetting, one must choose $\alpha_1 > \alpha_L$ (or, more precisely, $\alpha_S < \alpha_L < \alpha_1$). To render it non-wetting, one must choose $\alpha_1 < \alpha_L$ ($\alpha_1 < \alpha_L < \alpha_S$). However, the thin film coating does not entirely mask the properties of the original substrate.

4.2.4   Other Long-Range Forces

Electrostatic Interactions in Water

Quite often, a solid substrate exposed to water is apt to capture electrical charges. For example, $SiOH$ groups ubiquitous on the free surface of silica or glass become ionized (at least in a certain pH range) in accordance with the reaction $SiOH \rightarrow SiO^- + H^+$. This reaction gives rise to double layers. In a neutral pH environment ($\sim$7), most ordinary oxides create a negative surface surrounded by positive ions. This produces a first layer in the vicinity of the solid. There is also a second one at the free surface of water. If both surfaces carry charges of the same sign, the result will be a force of repulsion [$P(e) > 0$] whose range is the Debye-Hückel screening length κ_D^{-1}. In the case of monovalent ions, κ_D is given by

$$\kappa_D^2 = \frac{2ne^2}{\varepsilon_0 \varepsilon kT} \tag{4.27}$$

where n is the number of salt molecules (e.g., NaCl) per unit volume, e is the electronic charge, ε is the relative dielectric constant of water ($\sim$80), and ε_0 is the dielectric constant of vacuum. It is convenient to introduce Bjerrum's length, defined by $l_B = \frac{e^2}{4\pi\varepsilon_0\varepsilon_r kT}$ (equal to 7 Å for water), which leads to $\kappa_D^2 = 8\pi n l_B$. Typically, for 0.1 mole of salt per liter of water, $\kappa_D^{-1} = 1$ nanometer. For more details on these forces, consult the literature.[3]

Case of Flexible Polymer Solutions

Two very different situations arise here:

1. When the solid/liquid and liquid/air surfaces have a stronger affinity toward the polymer than toward the solvent, a *diffuse adsorbed* layer is formed. The size of this layer is related to the radius of the polymer coils and is typically 100 Å for a molecular weight of 10^6. Here again, the result is a long-range interaction that is (most often) repulsive.[4]

2. When the surfaces have a stronger affinity toward the solvent than toward the polymer, the polymer avoids the vicinity of the surfaces. As a result, a *depletion* layer forms, the thickness ξ of which depends on the concentration of the polymer.[4] Here again, depletion layers can be coupled and can generate an interaction of attraction [$\Pi(e) < 0$]. The dependence of $P(e)$ matches that described in Figure 4.4b and the resulting films are unstable. But the final state here is not a dry solid. Rather, it is a film of pure solvent in equilibrium with a drop of solute.[5]

4.3   Some Manifestations of Long-Range Forces

4.3.1   Films on Slightly Rough Substrates: The Healing Length

The situation considered here is depicted in Figure 4.7, where a substrate is characterized by small height variations $\zeta(x, y)$ about a horizontal plane $z = 0$. A liquid film is deposited on the substrate. The film is assumed to be totally wetting and in a stable regime ($e > e_c$). The question we address

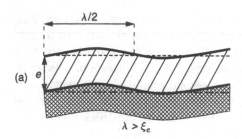

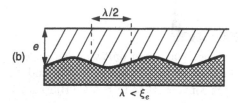

FIGURE 4.7. Thin film on a rough substrate. The wavelength λ is the characteristic pitch of the roughness to be compared with the healing length ξ_e.

now is the following: Does the surface of the liquid film replicate the rough profile of the substrate by keeping its thickness constant, or does it "smooth out" the original roughness?

We describe the modulation of the free surface by a displacement $u(x, y)$. The local thickness of the film is $e + u - \zeta$, and the energy W has the form:

$$W = \int dx\, dy \left\{ P(e + u - \zeta) + \frac{1}{2}\gamma \cdot (\nabla u)^2 \right\} \tag{4.28}$$

where the second term describes changes in the liquid/air surface area.

For small deformations, we expand $P(e + u - \zeta)$ out to second order in $(u - \zeta)$. The zero-order term gives a constant. The first-order term drops out since the spatial average of $(u - \zeta)$ is zero if the volume of the liquid is conserved. The second-order term yields the following equation for the optimum profile:

$$(u - \zeta)\ddot{P}(e) - \gamma\nabla^2 u = 0 \tag{4.29}$$

which can be rewritten as

$$-\xi_e^2 \nabla^2 u + u = \zeta. \tag{4.30}$$

The quantity ξ_e is a length given by

$$\xi_e = \left[\frac{\gamma}{\ddot{P}(e)}\right]^{1/2} \tag{4.31}$$

It is referred to as the *healing length* of the film.[6]

- If the variations of ζ are slow on the scale of ξ_e, so are the variations of u. We can then neglect the $\nabla^2 u$ term in equation (4.30). Given these assumptions, the result is $u = \zeta$, which means that the film retains a constant thickness and the free surface adopts the same rough profile as that of the starting substrate.
- If, on the contrary, the variations of ζ take place on a finer scale (less than ξ_e), then the dominant contribution to the energy is the capillary term and equation (4.29) is essentially satisfied with a flat profile $u = 0$. In this case, the initial surface roughness is smoothed out at the free interface.

These properties have a direct bearing on varnishes used as a finish on rough surfaces (such as wood or plastics). If the refractive index of the substrate is not too different from that of the varnish, it is possible to achieve a good surface polish provided that ξ_e is greater than the horizontal pitch of the surface roughness.

Example. If for the thicknesses of interest the energy $P(e)$ is dominated by a van der Waals term [equation (4.19)] with a positive Hamaker constant, we have

$$\ddot{P}(e) = \frac{A}{2\pi e^4} \tag{4.32}$$

and

$$\xi_e = \left(\frac{2\pi\gamma}{A}\right)^{1/2} e^2 \approx \frac{e^2}{a} \tag{4.33}$$

where a is a molecular length.

The conclusion is that ξ_e increases rapidly with the film thickness. It should be kept in mind, however, that equation (4.33) holds only to the extent that $e < 1,000$ Å (upper range of van der Waals forces).

4.3.2 Fine Structure of the Triple Line

In a partial wetting regime and in the immediate vicinity of the triple line, the profile of the liquid can deviate from the constant slope ($\tan \theta_E$) described by Young's equation. A legitimate question to ask then is whether long-range forces can cause visible deformations of the profile. The relevant spatial scale is typically a few nanometers, which is a challenge to observe.

A discussion of such profiles is particularly interesting when the contact angle θ_E is small.

- First, the relevant spatial scale then becomes somewhat larger.
- Second, calculations simplify because a liquid wedge can be treated locally as a film, which enables us to use the function $P(e)$. If $u(x)$ is the function describing the profile, the energy is still given (in the limit of

small slopes) by equation (4.28) after letting $\zeta = 0$ for a smooth surface. The equilibrium equation reads

$$\gamma\frac{d^2u}{dx^2} + \Pi(u) = 0. \tag{4.34}$$

This equation expresses the fact that the Laplace pressure and the disjoining pressure are in equilibrium. A first integration gives [see equation (1.34)]:

$$\frac{1}{2}\gamma\left(\frac{du}{dx}\right)^2 - P(u) = \frac{1}{2}\gamma\theta_E^2. \tag{4.35}$$

The integration constant $\gamma\theta_e^2/2$ ensures that at large distances ($u \to \infty$) we end up with $du/dx = \theta_e$, which is consistent with $P(u) \to 0$.

With this formula, we can determine the function $P(u)$—as long as we know the profile of the wedge—by measuring the local slope du/dx.

We begin by discussing the profile resulting from the involvement of pure van der Waals forces. To do so, we use equation (4.19), recast in the form

$$P(u) = \frac{1}{2}\gamma\frac{a^2}{u^2} \tag{4.36}$$

where we have chosen a Hamaker constant $A > 0$. The molecular length a is

$$a^2 = \frac{A}{6\pi\gamma}. \tag{4.37}$$

Integrating equation (4.35) once more yields

$$u^2 = \theta_E^2 x^2 - \frac{a^2}{\theta_E^2}. \tag{4.38}$$

The profile is hyperbolic in shape, as illustrated in Figure 4.8. The hyperbola merges with its asymptote at a height of the order of a/θ_E. While the molecular length a (~ 1 Å) is difficult to observe, the height a/θ_E is readily measurable when θ_E is small.

Admittedly, the preceding calculation is unrealistic for small values of u because the expression for $P(u)$ diverges when $u \to 0$. In actuality, $P(u \to 0)$ remains finite and is equal to the spreading parameter $S = -\gamma\theta_E^2/2$. In accordance with equation (4.35), the slope vanishes ($du/dx = 0$) at the triple line. The correction is shown as a continuous line in Figure 4.8. In fairness, though, such distinctions affect regions of size $\sim a$ that are too small anyway for the calculation to be meaningful.

For more detailed discussions of the local structure on a nanometer scale and of corresponding measurements by atomic force microscopy, the reader may want to consult the literature.[7-9]

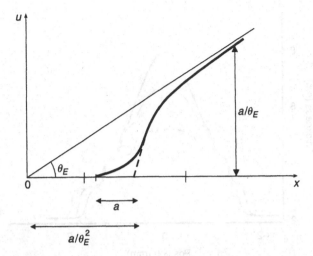

FIGURE 4.8. Profile in the vicinity of a triple line. The slope is zero where the profile merges with the solid substrate. The dotted line corresponds to a simplified form of the energy $[P(e) \sim 1/e^2$ for all values of $e]$.

4.4 Stratified Film

When certain molecular fluids spread on a substrate in a total wetting regime, they exhibit a structure resembling an Aztec pyramid (Figure 4.9) easily observable by ellipsometry.[10]

In the vicinity of solid surfaces, liquids tend to organize themselves in layers. The effect has been reproduced in numerous simulations involving a simple liquid in the presence of a wall.

To some extent, it is possible to analyze this property in terms of the quantity $P(e)$, which expresses the energy of a film as a function of its thickness. Suppose that the graph of $P(e)$ oscillates and displays at least a few sharp minima, as shown in Figure 4.10. When drawing a common tangent to the curve, one can envision successive layers $(1, 2, \ldots, n)$ with well-defined thicknesses, with the possibility that any two of them $(n, n+1)$ can coexist in equilibrium. The slope of the common tangent at n and $n+1$ is denoted $-\Pi_n$. In order for all the layers to be visible, the envelope defined by the series of common tangents must be convex. In other words, Π_n must decrease with increasing n, as shown in Figure 4.10. If the thickness is slightly larger than that of the n_{th} layer, regions with n and $n+1$ layers will coexist.

In practice, the most common outcome is the formation of two or three layers, particularly with fluids made of globular molecules. Still, even some polymers (such as silicones) show hints of incipient stratification. Not surprisingly, the tendency for stratification is particularly pronounced with liquids near a smectic state (which tend to pile up naturally in fluid layers).

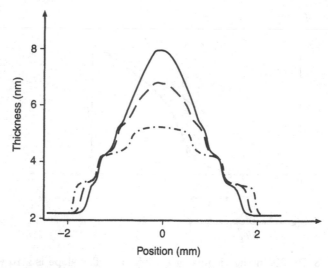

FIGURE 4.9. Profile of a drop of tetrasiloxane (whose molecule is rather small and spherical) deposited on a solid substrate. It exhibits terraces correspondingly roughly to the diameter of the molecule (courtesy F. Heslot et al.).[10]

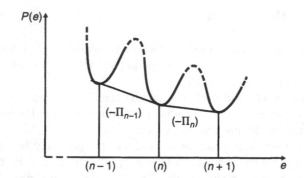

FIGURE 4.10. Energy $P(e)$ leading to a stratified film.

References

[1] A. Sheludko, *Adv. Colloid Interface Sci.* **1**, 391 (1967).

[2] K. Ragil, J. Meunier, D. Broseta, and D. Bonn, *Phys. Rev. Lett.* **77**, 1532 (1996).

[3] J. Israelashvili, *Intermolecular and Surfaces Forces* (New York: Academic Press, 1985).

[4] P. G. de Gennes, *Adv. Colloid Interface Sci.* **27**, 189 (1987).

[5] F. Rondecave and F. Brochard-Wyart, *Macromolecules* **31**, 9305 (1998)

[6] D. Andelman, J. F. Joanny, and M. Robbins, *Europhys. Lett.* **7**, 731 (1988).

[7] T. Pompe, A. Fery, and S. Herminghaus, Apparent and Microscopic Contact Angles, (J. Drelich, J. S. Laskowski, and L. K. Mittal, eds.) (Utrecht: VSP Editions, 2000), p. 3.

[8] A. Fery, T. Pompe, and S. Herminghaus, *ibid.*, p. 12.

[9] C. Bauer and S. Dietrich, *Europhys. J. B* **10**, 767 (1999).

[10] F. Heslot, N. Fraysse, and A. M. Cazabat, *Nature* **338**, 640 (1989).

5

Hydrodynamics of Interfaces: Thin Films, Waves, and Ripples

5.1 Mechanics of Films: The Lubrication Approximation

The dynamical properties of objects in motion are described by Newton's law, which states that the sum of all forces applied to an object is equal to its mass times its acceleration. The first difficulty one runs into when dealing with liquids is that different elements of volume making up the liquid can move independently of one another. This forces us it to work with small elements of volume, rather than with a rigid body moving as a whole. Under these circumstances, the fundamental equation reads

$$\vec{f} = \rho \frac{d\vec{v}}{dt} \tag{5.1}$$

where ρ is the volumetric mass of the liquid, $\vec{v}$ is the velocity of an element of volume, and $\vec{f}$ is the force acting on that element of volume. The force is generally the sum of a driving term (due to gravity, for instance) and a viscous term $\vec{f}_\eta$ that opposes any movement.

The above equation, together with a condition related to the incompressibility of the liquid ($\vec{\nabla}\vec{v} = 0$), constitutes what is known as the Navier-Stokes equation. It is normally extremely difficult to solve for three reasons:

1. It is a vector equation, and in three-dimensional space, three separate equations must be solved in order to determine the velocity vector,

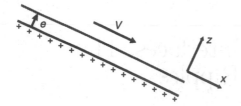

FIGURE 5.1. Thin liquid film flowing down an incline.

2. It is a non-linear equation because of the presence of an inertial term. For a flow characterized with an average velocity V over a characteristic length L, the inertial term is $\rho V^2/L$, which is quadratic in V,

3. It is a second-order equation, which requires two successive integrations with respect to the spatial coordinates. The reason is that the force $\vec{f}$ generally includes a viscous term that reads $\vec{f_\eta} = \eta\nabla^2\vec{v}$, where η is the dynamical viscosity of the fluid and ∇^2 is the Laplace operator.

For our purposes the first two difficulties can often be circumvented by adopting the so-called *lubrication approximation*. The rationale behind this simplification is the fact that flows generally have one dimension (transverse) much smaller than the others. Consider, for example, a thin liquid film flowing down an incline under the effect of gravity (Figure 5.1).

Several points are noteworthy. First, the velocity vector at every location of the liquid points mainly in the x-direction. As a result, it is legitimate to reduce the Navier-Stokes equation to its v-component along the direction of the flow, thereby replacing the original three-dimensional vector equation with a much friendlier scalar counterpart. Second, the boundary condition at the solid/liquid interface demands that the velocity be continuous. Since the solid is fixed, it stands to reason that the liquid in its immediate vicinity must stand still as well. This restriction has two consequences:

• If the film is thin, its velocity is bound to be slow. As a result, the inertial term (which, as we have just pointed out, is proportional to V^2) can safely be neglected when compared to the viscous term (which is proportional to V). The ratio between these two terms is known as the Reynolds number. Whenever this number is much smaller than 1, as is the case for most flowing thin films, it is appropriate to neglect the inertial term in the Navier-Stokes equation.

• Notwithstanding the zero-velocity at the solid substrate, the film does support an overall flow. This implies a velocity gradient in the direction perpendicular to the flow. Therefore, there is a decoupling between the driving force, which fixes the direction of the flow, and the viscous force, which is related to the velocity gradient.

We still have to specify just what the driving force is. In the example of Figure 5.1, it was gravity. More generally, a force responsible for moving a fluid horizontally (sometimes even pushing it back up a slope) is a pressure gradient. If two regions of a fluid are subjected to different pressures (the pressure difference is denoted Δp), a net transfer of fluid will take place from the high-pressure region to the low-pressure region. If L is the distance between these two regions, the force per unit volume responsible for the transport is the pressure gradient set up between the two regions over that distance. In other words, $-\partial p/\partial x = -\Delta p/L$ (the minus sign comes about because the pressure decreases in the direction of the flow). If desired, one might even include gravity in the pressure gradient by introducing the hydrostatic pressure.

Under these conditions, $\vec{f} = 0$ since the inertial term has been neglected, and the Navier-Stokes equation simply reduces to

$$-\frac{\partial p}{\partial x} + \eta \frac{\partial^2 v}{\partial z^2} = 0. \tag{5.2}$$

At this point, we are faced with two possible strategies.

The first is to solve the above equation, which requires two successive integrations with respect to the variable z. Since the driving force is independent of z, the double integration leads inevitably to a velocity profile $v(z)$ that is *parabolic* (such a dependence is known as a Poiseuille profile). To specify the profile completely, it is necessary to invoke two boundary conditions. We have already discussed the first, which dictates that no fluid motion be allowed at the solid wall. The second condition is for the stress to be continuous. The free interface (between the liquid and its vapor in the example of Figure 5.1) cannot withstand any viscous stress (that is to say, a force per unit surface). Indeed, there is nothing beyond the free surface to counterbalance such a force, given the fact that the viscosity of air dwarfs that of even the least viscous fluid by a factor of about a hundred. Since the viscous stress is given by $\eta \partial v/\partial z$, it follows that the velocity gradient must vanish at that interface. Stated differently, the velocity must reach its maximum value there. This set of two conditions ($v = 0$ at $z = 0$, and $\partial v/\partial z = 0$ at $z = e$) allows us to work out the profile at any point within the fluid. The reader is encouraged to do the math as an exercise. She or he may also show that there does exist a viscous stress at the solid/liquid interface. There is nothing alarming about this, since the solid is quite capable of withstanding such a stress thanks to its elasticity. We have schematically illustrated the velocity profile in Figure 5.2, where we have assumed that the film flows as a result of a pressure difference $P_+ - P_-$ between upstream and downstream regions.

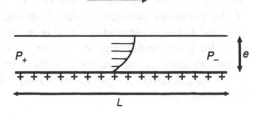

FIGURE 5.2. Velocity profile in a flowing thin film.

The second approach is to write all these results directly in the form of a scaling law. Since the velocity V varies essentially over a distance e, the viscous term has an order of magnitude $\eta V/e^2$, where V now refers to the *average* velocity of the fluid. Equation (5.2) then becomes

$$V \propto \frac{e^2}{\eta}\frac{\Delta p}{L}. \tag{5.3}$$

The flow rate Q of the liquid (expressed per unit film width) is simply eV. If f denotes the force per unit volume in the x-direction that sets the liquid in motion (ordinarily due to either gravity or a pressure gradient $-\partial p/\partial x$), the flow rate can be written as

$$Q \propto \frac{e^3}{\eta}f \tag{5.4}$$

These equations are strikingly simple. Both the liquid's velocity and its flow rate are proportional to the driving force. This property is known as *Poiseuille's law*. One can define a hydrodynamic conductivity, which is the ratio Q/f and is inversely proportional to the viscosity η. The viscosity is an intrinsic property of the liquid (for a given force, the greater the viscosity, the lower the flow rate). However, this conductivity depends mainly on the thickness of the liquid film, actually varying as the cube of that thickness. A film twice as thick will support eight times the hydrodynamic flow. This is an important characteristic that comes up in most interface hydrodynamics problems, which are generally dominated by viscous friction.

> **Problem:** Using equation (5.2) and its two boundary conditions to determine the velocity profile, calculate the flow rate of a liquid and its average velocity. Show that the numerical coefficient in equations (5.3) and (5.4) is equal to 1/3 and that the velocity at the free surface equals 3/2 times the average velocity.

Knowing the scaling laws [equations (5.3) and (5.4)] enables us to handle most viscous interface hydrodynamics problems dimensionally. Even though that approach may overlook the details of flows, the scaling laws provide us with a quick picture of the behavior of such flows (in terms of the pertinent parameters and how important quantities vary as functions of these parameters). In this sense, these laws provide a valuable path for problem solving. Whenever questions arise that involve cumbersome calculations, we will not hesitate to resort to this approach, all the while encouraging the reader to consult original papers for the details of the calculations.

5.2 Dynamics of Thin Films

The dynamics of thin liquid films are often dominated by viscous friction. As such, the lubrication approximation is entirely appropriate. In this section, we discuss four examples in order of increasing difficulty. They are the thinning of a vertical liquid film, the levelling of a wavy film, and two examples of interfacial instability (suspended film and liquid cylinder). We restrict our attention here to films of thickness $e > 100$ nm, for which the long-range interactions described in chapter 4 can be neglected $[P(e) \to 0]$.

5.2.1 Thinning of a Vertical Film

Consider a thin liquid film placed on a vertical substrate. Because of gravity, this film will flow and thin down as time progresses. We denote by $x = 0$ the position of the top of the film and by e_0 its initial thickness (Figure 5.3).

Since the force of gravity per unit volume is ρg (ρ is the density of the liquid and g is the acceleration of gravity), the flow rate of the liquid can be written as [see equation (5.4) and the subsequent problem]:

$$Q = \frac{e^3}{3\eta} \rho g \tag{5.5}$$

where e is the thickness of the film. Since the film thickness decreases from the top down, e is expected a priori to be a function of position x and time t. Assuming that the film thins down at a rate $\partial e/\partial t$, the flow rates at positions x and $x + dx$ differ by an amount $(\partial e/\partial t)\, dx$. Therefore, the conservation of volume is expressed as

$$\frac{\partial Q}{\partial x} = -\frac{\partial e}{\partial t}. \tag{5.6}$$

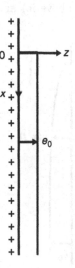

FIGURE 5.3. Thin liquid film on a vertical wall.

From there we deduce an equation describing the evolution of the profile $e(x,t)$:

$$\frac{\rho g}{\eta}e^2\frac{\partial e}{\partial x} = -\frac{\partial e}{\partial t}. \qquad (5.7)$$

A first solution to this equation is $e = $ constant, which implies $e = e_0$. We next search for a second solution of the type $e(x,t) = f(x) \cdot g(t)$. This leads to a differential equation of the type $f \cdot f' = -g'/g^3$. The two sides must reduce to a constant since the left-hand side depends on x only, while the right-hand-side depends on t only. The final result is known as Reynolds' thinning law:[1]

$$e(x,t) = \sqrt{\frac{\eta x}{\rho g t}}. \qquad (5.8)$$

This solution does start with zero thickness at the top of the film ($x = 0$). It holds if the triple line remains pinned at that spot (see chapter 3). In this case, the profile of the film near the top is parabolic. In that region, the thickness (at a given x) decreases slowly (as $t^{-1/2}$). The solution is valid until $e = e_0$ [the other solution of equation (5.7)], which corresponds to a distance from the top $L(t) \approx \rho g e_0^2 t/\eta$. The length of the slimmed-down region progresses linearly in time. Its velocity is essentially fixed by the thickness of the film and can be quite low for thin films (it is equal to 1 mm/s for a 10 μm-thick film of water). Globally, the film has the profile schematically indicated in Figure 5.4.

5.2.2 Levelling of a Horizontal Film

Consider now a liquid film placed on a horizontal substrate. Suppose that the free surface has been perturbed, for instance, by exposure to an air flow (Figure 5.5). One might guess that this perturbation should disappear over time under the combined effects of gravity (the liquid is denser than the vapor above it) and surface tension (roughness increases the surface area

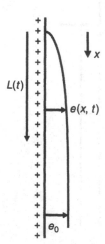

FIGURE 5.4. Profile of a vertical film thinning down under the influence of gravity.

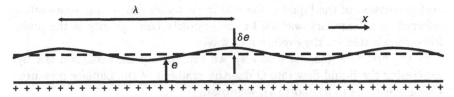

FIGURE 5.5. Ripples on the surface of a thin film.

of the liquid/air interface). Our intent is to figure out the time constant of this relaxation process by restricting ourselves for the sake of simplicity to the case of a one-dimensional roughness profile, with the surface shaped like a corrugated panel.

We write the equation of the free surface in the form

$$e = e_0 + \delta e \cos(qx) \tag{5.9}$$

where e_0 is the average thickness of the film, q is the wavevector of the disturbance ($\lambda = 2\pi/q$ is its wavelength), and δe is its amplitude. Moreover, we assume that $e \ll \lambda$ and that the viscous force dominates over the inertia of the fluid. Then the problem can be tackled in the lubrication approximation.

The first question we address concerns the distribution of pressure in the fluid. To start with, the hydrostatic pressure is modulated along the x-direction. It is higher underneath a crest than underneath a trough. As a result, the liquid will tend to flow from crests to troughs because of the weight of the liquid, as we have already pointed out earlier. Additionally, the Laplace pressure, which is related to the curvature of the free surface, is modulated as well. Here again, the pressure is higher under a crest (with the curvature pointing toward the liquid) than it is under a trough (where the curvature points toward the air). Thus, whether it is of gravitational or capillary origin, the pressure is modulated in phase with the thickness. Flows will, therefore, be set up from crests to troughs. This will enable the surface to revert to a planar configuration.

The first step is to determine whether one of the two mechanisms at play (gravity and capillarity) might dominate over the other. The hydrostatic pressure difference between a crest (or a trough) and the average height of the film is $\rho g \delta e$. The Laplace pressure, on the other hand, is equal to the product of the surface tension γ of the liquid and the curvature C of the interface. For a weakly curved surface (implying $|de/dx| \ll 1$), $C \approx -d^2 e/dx^2$.

Under these conditions, the difference in curvature between a crest (or a trough) and the median line is of order of magnitude $\delta e/\lambda^2$. The ratio of the hydrostatic pressure to the Laplace pressure for the wave depicted in Figure 5.5 is $(\lambda \kappa)^2$, where κ^{-1} is the capillary length introduced earlier. The conclusion is that the way ripples or waves evolve is determined by the wavelength. If the ripples have a short wavelength ($\lambda < \kappa^{-1}$), the surface tension of the liquid is the dominant factor (such ripples are often referred to as capillary waves). In the opposite case, gravity is the main force responsible for the evolution of waves.

In the first case (capillary waves), we can specify the evolution law by expressing the liquid flow rate Q due the gradient of the Laplace pressure. Using equations (5.9) and (5.4), we obtain

$$Q \approx \frac{e_0^3}{\eta} \gamma q^3 \delta e \sin(qx) \tag{5.10}$$

where the film thickness has been approximated by its average value in the expression for the hydrodynamic "conductivity" (a legitimate step since $\delta e \ll e$). Making use of the volume conservation equation (5.6), we arrive at an expression describing the temporal evolution of the ripple amplitude δe:

$$\frac{d\delta e}{dt} = -\frac{\delta e}{\tau}$$

where the time constant τ follows the scaling law:

$$\tau \approx \frac{\eta \lambda^4}{\gamma e_0^3} \tag{5.11}$$

The exact forum for this expression is $\tau = \frac{3\eta}{\gamma e_0^3 q^4}$.

We see that a capillary wave in a viscous thin film relaxes exponentially as a function of time. The characteristic time of the relaxation process is a function of the liquid—it varies as the inverse of a characteristic velocity $V^* = \gamma/\eta$—and, more important, of the geometrical characteristics of the film (through λ and e_0). For $\lambda = 1$ mm, $e_0 = 100$ μm, and for an ordinary cooking oil ($\eta \sim 200$ mPa-s and $\gamma \sim 20$ mN/m, meaning $V^* = 10$ cm/s), we find a time constant of the order of 10 s.

In the case of waves with a large wavelength, which are dominated by gravity, we can use a very similar argument. The difference is that now the force in equation (5.4) is provided by the hydrostatic pressure gradient $\rho g \partial e / \partial x$ [the pressure field in the undulating film is $p = P_0 - \gamma e'' + \rho g(e - z)$, where P_0 is the external pressure and z is the altitude. In the gravity regime, $\frac{\partial p}{\partial x} = \rho g \frac{\partial e}{\partial x}$]. The liquid is driven by the crest-to-trough slope of the pressure field. Differentiating e in equation (5.9) with respect to x leads to $f = \rho g \delta e \cdot q \sin(qx)$. We find that the relaxation process is again exponential, with a time constant given here by

$$\tau \approx \frac{\eta \lambda^2}{\rho g e_0^3} \tag{5.12}$$

Inserting the exact coefficient, the result becomes $\tau = \frac{3\eta}{\rho g e_0^3 q^2}$.

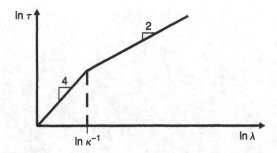

FIGURE 5.6. Relaxation time τ of a sinusoidal film as a function of the wavelength of the disturbance. For ripples ($\lambda < \kappa^{-1}$), τ increases as λ^4. For waves ($\lambda > \kappa^{-1}$), τ increases as λ^2.

For the same cooking oil as used in the previous example, a film with an average thickness of 100 µm and with $\lambda = 1$ cm has a time constant τ of the order of half an hour.

Finally, we can summarize this entire discussion by plotting the relaxation time τ as a function of the wavelength λ (Figure 5.6).

The graph shows two distinct branches in different wavelength ranges. The first branch corresponds to capillary relaxation [equation (5.11)], while the second corresponds to gravitational relaxation [equation (5.12)]. Our findings in terms of the relaxation time mirrors what we had concluded earlier in our discussion of forces: Capillarity dominates at wavelengths smaller than κ^{-1}.

By way of conclusion, we emphasize that this discussion of the levelling of waves is valid for thin films (such as a coat of paint, for instance) dominated by viscous friction. For very thick films, the analysis is different because of the role played by the inertia of the liquid. We shall discuss this case in section 5.5 of this chapter.

5.2.3 Rayleigh-Taylor Instability

The previous section dealt with the return to equilibrium of a thin film that had been perturbed. Conversely, it is possible for an interface to destabilize spontaneously under the influence of an external field.[2-8] The field can simply be gravitational.[7,8] Whereas gravity had a stabilizing influence in the previous section, things can be different for a suspended film (wet paint on a ceiling, for example). After a while, the film turns into an array of drops such as shown on the photograph in Figure 5.7. We can witness such clusters of drops hanging from the ceiling panel inside a refrigerator, where water vapor condenses.

We shall now discuss a suspended film and its instability (known as *Rayleigh-Taylor instability*). We shall first find what wavelengths turn out to be unstable and then go on to examine the dynamics of the phenomenon.

FIGURE 5.7. Drops hanging under a horizontal panel. The drops are formed either by condensation or by destabilization of a liquid film. Each drop is approximately 1 cm in size (courtesy Marc Fermigier).

We retain the notation introduced earlier in equation 5.9. The film thickness is modulated with an amplitude δe about the median value e_0 ($\delta e \ll e_0$) and a wavevector q. The wavelength $\lambda = 2\pi/q$ is assumed to be large compared to e_0. We also assume for simplicity that the deformation is a function of the coordinate x only (the corrugated panel model).

Two opposite effects act on the surface. Gravity tends to distort it by transporting dense liquid downward, while surface tension resists any tendency to increase the surface area. We evaluate the energy difference (per unit length in the direction perpendicular to Figure 5.5) between a wavy film and a flat one over a distance λ:

$$\Delta E = -\int_0^\lambda \frac{1}{2}\rho g(e - e_0)^2\, dx + \int_0^\lambda \gamma(ds - dx). \tag{5.13}$$

The surface term can be evaluated by calculating the curved length of the interface. In the limit of small deformations ($\delta e \ll \lambda$), the quantity ds can be expanded in the form $ds \approx dx[1 + (1/2)(de/dx)^2]$. Equation (5.13) can then be readily integrated to yield

$$\Delta E = \frac{1}{4}\gamma \delta e^2 \lambda \cdot (q^2 - \kappa^2). \tag{5.14}$$

The conclusion is that ΔE is negative when $q < \kappa$. Therefore, long wavelengths ($\lambda > 2\pi\kappa^{-1}$) will be unstable. Practically, the length $2\pi\kappa^{-1}$ is at least 1 cm. What this means is that if one turns (skillfully) an open test tube filled with water upside down, the content of the tube will not drain out. The largest wavelength that could develop is comparable to the diameter of the test tube (assume it to be equal to 1 cm), which is less than the smallest unstable wavelength ($2\pi\kappa^{-1}$).

There is an infinite number of unstable modes, namely, all those with a wavelength greater than $2\pi\kappa^{-1}$. However, as evidenced by the regularity of the array of drops in Figure 5.7, a single mode gets selected in practice. To understand why that is, we must understand the dynamics of the phenomenon. We do so by extending the argument used in the previous section.

Equation (5.4) gives the flow rate of a liquid that is due to a volumetric force ∇p. In the present case, the force is determined both by the weight of the liquid, which is $\rho g \partial e/\partial x$ per unit volume, and by the Laplace force $-\gamma\partial^3 e/\partial x^3$ induced by the surface curvature. The force, and hence the flow rate, can thus easily be deduced from the interface profile [equation (5.9)]. Invoking the flow rate equation (5.6) leads to an equation describing the evolution of the interface:

$$\frac{d\delta e}{dt} = \delta e \frac{\gamma e_0^3}{3\eta} q^2(\kappa^2 - q^2) \tag{5.15}$$

where we have assumed that friction takes place over a constant thickness e_0.

Equation (5.15) is linear and has a solution that is exponential in time. A disturbance of the initial interface (due for instance to thermal fluctuations) will be amplified if the quantity $(\kappa^2 - q^2)$ is positive, which confirms the wavelength criterion derived from equation (5.14). What is new here is that each unstable wavevector can be assigned its own characteristic time τ given by

$$\tau(q) = \frac{3\eta}{\gamma e_0^3} \frac{1}{q^2(\kappa^2 - q^2)}. \tag{5.16}$$

The dependence of τ on q is not monotonic. The time constant describing the rate of growth of an unstable mode diverges when $q \to 0$ (that is to say, for an infinite wavelength, which implies transport of matter over an infinite distance), and again when $q \to \kappa$ (in other words, just on the edge of instability). Between these two values, $\tau(q)$ goes through a minimum τ^*. This minimum time occurs when $q^* = \kappa/\sqrt{2}$ and is equal to

$$\tau^* = 12\frac{\gamma\eta}{\rho^2 g^2 e_0^3}. \tag{5.17}$$

As Rayleigh understood very well a long time ago, the particular wavelength $\lambda^* = 2\pi/q^*$ wins out for kinetic reasons. Of all the possible unstable wavelengths, it is the one that grows the fastest and overshadows all the others. That wavelength is given by

$$\lambda^* = 2\pi\sqrt{2}\kappa^{-1} \tag{5.18}$$

These main results can be derived more expediently from dimension arguments. Since the Rayleigh-Taylor instability results from the competing effects of gravity and surface tension, it is logical to expect that its wavelength should be controlled by the capillary length $\kappa^{-1} = \sqrt{\gamma/\rho g}$. The dynamics of the process is controlled by Poiseuilles' law, which, based on equation (5.3), is given dimensionally by $\lambda/\tau \propto (e^2/\eta)(\rho g e/\lambda)$ since matter is being transported over a distance of order λ along a slope of order e/λ. For $\lambda \approx \kappa^{-1}$, we recover equation (5.17) (to within a numerical coefficient). Finally, remember that we have assumed the flow to conform to Poiseuille's law, which implies that it is dominated by viscous friction.

5.2.4 Plateau-Rayleigh Instability

A second classic type of interfacial instability is met with liquid cylinders.[9] Consider, for instance, a fiber (a hair, with a typical diameter of 100 µm, is a good example) encased in a liquid sheath. Generally, such a film proves unstable. It develops a wavy profile (Figure 5.8).

The wave-like structure keeps on growing in amplitude until it eventually breaks up into a necklace of droplets (Figure 5.9).

The fact that the axial symmetry is retained is a sure sign that we are not dealing here with a new version of the Rayleigh-Taylor instability since as much liquid rises above its original level as dips below. More quantitatively, the pressure within the film includes both a hydrostatic contribution of order $\rho g R$ and a Laplace contribution of order γ/R (where $R = b + e_0$, with b denoting the radius of the fiber and e_0 denoting the initial thickness of

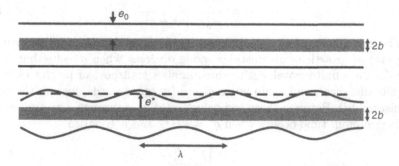

FIGURE 5.8. Liquid film on a fiber. The film destabilizes spontaneously.

FIGURE 5.9. Necklace of droplets resulting from the instability of a liquid film on a fiber 100 µm in radius (courtesy Elise Lorenceau).

the liquid sheath). A parameter Bo, known as the *Bond number*, compares these two pressures. It can be written as

$$Bo = \kappa^2 R^2. \tag{5.19}$$

For a liquid cylinder ≈ 100 μm in diameter, the Bond number is typically 0.01—a clear indication that gravity is negligible. The present instability is controlled exclusively by surface tension, which may sound surprising to some since we have just argued that surface tension has a stabilizing effect in the Rayleigh-Taylor instability. However, the phenomenon we are dealing with here is intimately related to the cylindrical geometry of the system, as mentioned in section 1.1.5. The wavy surface area shown in Figure 5.8 is actually *less than that of the original cylinder*, at least if the wavelength is sufficiently large, as we will show shortly.

Let $e = e^* + \delta e \cos(qx)$ represent the modulation of the film thickness. Note that the conservation of volume imposes that e^* be different from e_0. In fact, $e^* < e_0$, which is a first hint that the film's modulation serves to lower its surface area. We find

$$e^* = e_0 - \frac{\delta e^2}{4R}. \tag{5.20}$$

▎ *Problem:* Prove equation (5.20).

The second step in the argument is to duplicate the analysis we went through in the previous section by writing the energy difference ΔE between the modulated surface in Figure 5.8 and the straight cylinder from which it originated. Since the surface energy is the only contribution in the present case, the energy difference over a wavelength now reads

$$\Delta E = \int_0^\lambda 2\pi(b + e)\gamma \, ds - 2\pi(b + e_0)\gamma\lambda.$$

Assuming again that the surface profile is weakly modulated $\{ds \approx dx[1 + (1/2)(de/dx)^2]\}$, we find

$$\Delta E = \frac{1}{4}\gamma\frac{\delta e^2}{R}2\pi\lambda(q^2 R^2 - 1). \tag{5.21}$$

This result shows clearly that the energy is lowered ($\Delta E < 0$) if the wavevector verifies the inequality $q(b + e_0) < 1$, in other words, if the wavelength is greater than the perimeter of the original cylinder:

$$\lambda > 2\pi R. \tag{5.22}$$

As an interesting footnote, this inequality was first established experimentally. Around 1870, Plateau was busy immersing drops of oil in water containing some alcohol (to adjust the density). He proceeded to deform them into cylinders using spatulas which he manipulated laterally. He noticed that the cylinder became unstable when its length was sufficiently large. He determined that the instability occurred when the ratio of the cylinder's length to its diameter was between 3.13 and 3.18. In doing so, he became the first to discover the shortest unstable wavelength, as written in equation (5.22).

As in the previous section, dynamical properties will pick the winner from among an infinity of unstable wavelengths.[10] In the interest of simplicity, we choose to work in the thin film limit $e_0 \ll b$. The two fundamental equations are still Poiseuille's equation (we assume that the viscosity dominates over inertia—an eminently reasonable assumption for thin films) and the flow rate equations (5.5) and (5.6). The only tricky point here is how to write the pressure gradient responsible for setting up the flow. At any point of the interface, we have two curvatures to deal with in the present case—one in the plane of Figure 5.7, and another (in a perpendicular plane) associated with the fiber. Under these conditions, the curvature C in the limit of small slopes is given by

$$C \approx \frac{1}{b+e} - \frac{d^2e}{dx^2}.$$

It is the gradient dC/dx of this curvature that enters equations (5.5) and (5.6). The calculation yields the following equation describing the evolution of the perturbation:

$$\frac{d\delta e}{dt} = \delta e \frac{\gamma e_0^3}{3\eta b^2} q^2 (1 - q^2 b^2). \tag{5.23}$$

Again, we encounter our previous instability criterion. A perturbation will grow $(d\delta e/dt > 0)$ provided that $qb < 1$ [that is the same as condition (5.22) with $e_0 \ll b$]. To each wavelength we can associate a time constant $\tau(q)$, which varies as the inverse of $q^2(1 - q^2 b^2)$. This time constant is minimum for a particular wavelength λ^*:

$$\lambda^* = 2\pi\sqrt{2}b \tag{5.24}$$

Equation (5.24) describes the wavelength that the system will select, which explains the regularity of the string shown in Figure 5.9. The characteristic growth time of the instability is

$$\tau^* = 12\frac{\eta b^4}{\gamma e_0^3}. \tag{5.25}$$

This time can be quite long for a thin film. For example, for $e_0 = 1$ μm, $b = 100$ μm, using an ordinary cooking oil ($\gamma/\eta = 0.1$ m/s), the result is several hours. Note, however, that the result is extremely sensitive to the actual values of the relevant parameters. For instance, for $e_0 = 10$ μm on the same fiber, the time drops down to a mere 10 s.

Some concluding remarks are in order.

1. The final state of the instability sometimes manifests itself as a string of large drops and small ones ("satellites"). The satellites originate from a secondary instability of the liquid sheath as two adjacent drops are in the process of forming. When we deal with instabilities of fluid streams (particularly with viscous liquids), it is sometimes possible to observe a whole hierarchy of satellites, the appearance of which is considered a nuisance in many applications (for example, in ink jet printers, where the phenomenon has a disturbing tendency to degrade the resolution of the printed text).

2. Another important practical question is to figure out whether a microscopic liquid film can continue to exist between adjacent drops. This is likely to be the case when long-range forces favor thicker films. For instance, when the film is fairly thin, van der Waals forces oppose any further thinning that would normally occur for a Plateau-Rayleigh instability. A microscopic film can then coexist with drops. Its equilibrium thickness can be calculated easily by assuming equal pressures in the film and in the drops.[11] The pressure in the drops is approximately the same as the atmospheric pressure, while in the film it has a Laplace contribution (of order γ/b since $e \ll b$) as well as a disjoining contribution [equal to $A/6\pi e^3$ in accordance with equation (4.21)]. The thickness of the microscopic film thus comes out to be $e = a^{2/3}b^{1/3}$, where $a = (A/6\pi\gamma)^{1/2}$. The dimension a is microscopic (of the order of 1 Angström), and e turns out to have a typical value of 10 nm.

3. The Plateau-Rayleigh instability must also occur in cylindrical films coating the inside of a capillary tube. All the arguments developed for fibers apply here as well, including the comments just made concerning the coexistence of drops and film. The only significant difference here is the possibility that drops might form lenses stretching all the way across a tube when the film is thick enough.

4. In his original calculations, Rayleigh was interested primarily in the instability of a liquid jet of radius R, which breaks up into droplets, as Savart demonstrated early on.[12] While the static analysis remains unchanged [equations (5.20), (5.21), and (5.22) continue to hold], things are obviously different in terms of the dynamics. Dispensing with the fiber reduces the sources of viscous friction to a considerable degree. The characteristic time of the instability is now determined by the equilibrium between the inertia term ($\rho R/\tau^2$) and the capillary term (λ/R^2). This leads to $\tau \approx (\rho R^3/\gamma)^{1/2}$. For a millimeter-size water stream, this time turns out to be quite small (about 3 ms), which is significantly shorter than the free fall time over a height of 50 cm (about 30 ms). This is

consistent with what anyone can observe when turning a water faucet on. The water emerges as a laminar flow from the faucet but gets completely fragmented by the time it splashes against the sink (the best way to verify this fact is to place your hand in the lower part of the stream and clearly feel individual drops striking your skin). There also exists a viscous version of the calculation applicable to the case of highly viscous fluids (or fluids surrounded by another, highly viscous, fluid).[2,13] This can be used to advantage for measuring the surface tension of molten polymers, which can be deduced from the dynamical properties of the instability.

5.3   Forced Wetting

When trying to coat a solid with a liquid, a time-honored practice is to help the process along with some kind of movement. We all know that a paintbrush wiped against a wall or a piece of paper leaves a trace of liquid in its wake. Likewise, most industrial processes intended to deposit a fluid layer rely on a relative motion between solid and fluid. As an example, to apply a photographic emulsion on a suitable supporting medium, the standard method is to immerse a plastic strip in an emulsion bath and take it out gently so that the liquid "sticks" to the plastic. In any event, it is important to know the parameters that determine the thickness of the coating, which must be controlled precisely in many cases. This section is devoted to this particular problem. We shall confine our attention here to wetting liquids (the case of partial wetting will be discussed in chapter 6).

5.3.1   The Landau-Levich-Derjaguin Model (and Variant Thereof)

Consider first a solid plate partially immersed in a liquid, as we are about to pull it out. If the liquid is wetting ($S > 0$), the surface of the bath connects with the solid with a zero angle, which creates a meniscus in the vicinity of the line of contact. We have already discussed in chapter 2 some of the characteristics of this meniscus, including its height, which is of the order of the capillary length κ^{-1} [equation (2.20)].

Figure 5.10 depicts what happens when the plate is being slowly drawn out of the bath. The upper part of the meniscus (shown as a dotted line) finds itself perturbed by the liquid film dragged along by the plate. The junction between the static meniscus and the film being dragged along is referred to as the *dynamical meniscus*, and its length—unknown as of yet—is denoted l.

If the film is being pulled slowly enough, the film it drags along is thin since it would not exist at all absent the drag effect (ignoring the possible

FIGURE 5.10. Plate being pulled out of a pool of wetting liquid. The plate drags a liquid film along with it.

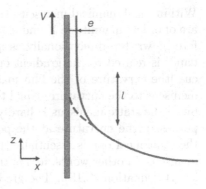

existence of a microscopic wetting film). Therefore, the key role is played by the interfaces:

1. The solid/liquid interface, because the associated boundary condition is actually responsible for the liquid coating. By virtue of its viscosity, the liquid in the vicinity of the solid moves at the same velocity as the solid and is being dragged by it,
2. The liquid/vapor interface, which is being distorted by the film despite the opposing action of the surface tension of the liquid. Of course, gravity causes the film to flow downwards as well and is therefore also opposed to its movement, although we will see that, when the so-called capillary number is small, the contribution of this force is negligible when compared to that of the surface tension.

In short, viscosity and capillarity play opposing roles. The number that compares these two forces (written per unit length) is called the *capillary number*, denoted Ca:

$$Ca = \frac{\eta V}{\gamma} \qquad (5.26)$$

This number is dimensionless, unlike the thickness itself. Therefore, there must be a normalizing length that is naturally associated with the meniscus from which the film originates. It happens to be the capillary length. Accordingly, the thickness is expected to be of the form $e = \kappa^{-1} f(Ca)$.

What follows is the argument originally proposed by Landau, Levich, and Derjaguin (LLD for short) in 1942–1943 to evaluate the thickness of the coating and to specify the functional form of $f(Ca)$. We present the LLD calculation in a simplified form. Here again the interested reader is encouraged to read the original papers, which qualify as veritable jewels in the field of interfacial hydrodynamics.[14,15]

In steady state and for low Reynolds numbers (when viscosity dominates over inertia), we can legitimately work in the lubrication approximation.

Within the dynamical meniscus, the velocity of the liquid and its thickness are of orders magnitude V and e, respectively. The viscous force can therefore be written dimensionally as $\eta V/e^2$. The capillary force, on the other hand, is related to the gradient of the curvature of the dynamical meniscus [the curvature of the film makes a transition from that of the static meniscus to the curvature (equal to zero) of the film being dragged along]. Since the static meniscus is hardly perturbed at all by the flow (LLD hypothesis), the curvature at the point where the static meniscus matches the dynamical one is essentially that which exists at the top of the static meniscus. In other words, it is of the order of κ, the inverse of the capillary length [equation (2.15)]. The gradient of the Laplace pressure associated with this curvature gradient is then equal to $\gamma\kappa/l$.

Equilibrium between these two forces is written dimensionally as

$$\frac{\eta V}{e^2} \approx \frac{\gamma\kappa}{l}. \tag{5.27}$$

This equation has two unknowns (e and l), and finding the solution requires an additional equation. The second equation derived by LLD expresses the matching between the static and the dynamical meniscus. We have already pointed out that the order of magnitude of the curvature is κ. The dynamical meniscus is nearly flat since it merges with a planar film. Therefore, its curvature can be taken to be equal to the second derivative of the profile $e(z)$, which is dimensionally given by e/l^2. These two curvatures have the same sign (both regions are in an underpressure condition with respect to the atmosphere). Setting them equal leads to

$$\frac{e}{l^2} \approx \kappa. \tag{5.28}$$

The extent of the dynamical meniscus therefore turns out to be the geometric mean of the other two relevant lengths in the problem, namely, the thickness e and the capillary length κ^{-1}:

$$l \propto \sqrt{e\kappa^{-1}} \tag{5.29}$$

which "happily" reduces to zero when the thickness e vanishes, in other words, when one stops pulling the plate. Equation (5.29) enables us to eliminate l in equation (5.27), which yields the law describing the thickness of the coating (known as the *LLD law*):

$$e \propto \kappa^{-1} Ca^{2/3} \tag{5.30}$$

With the help of equations (5.29) and (5.30), one can also derive the length of the dynamical meniscus:

$$l \propto \kappa^{-1} Ca^{1/3}. \tag{5.31}$$

In their rigorous calculations, LLD were able to evaluate the numerical coefficient in equation (5.30). Its turns out to be equal to 0.94. The calculation is done by writing the matching condition between the two menisci based on the mathematical expression for the profile of the free interface. At any rate, the underlying principle remains the same as before in the sense that the curvatures of both menisci are evaluated asymptotically. The curvature of the static meniscus is calculated at its summit (even though the actual matching may occur slightly above), while that of the dynamical meniscus is calculated in the limit when its thickness is large in comparison with e (it can be shown by numerical calculation that the matching takes place over a thickness of the order of $10e$). A prerequisite for the previous argument to be valid is that the static and dynamical menisci differ only by a slight perturbation, which can be written in terms of their respective lengths $l \ll \kappa^{-1}$. Based on equation (5.31), we conclude that the LLD result holds as long as

$$Ca \ll 1 \qquad\qquad (5.32)$$

which amounts to restricting the velocity to low values (more rigorously, the requirement is $Ca^{1/3} \ll 1$, which is more restrictive). In practice, the LLD law will be valid for capillary numbers less than 10^{-3}. Another requirement is that the flow be nearly parallel to the plate, so that the velocity vector can be taken equal to it scalar component in the direction of the plate. This imposes $e \ll l$. In light of equations (5.29) and (5.30), this condition is completely equivalent to the one spelled out previously.

It becomes clear that the LLD model adequately describes situations in which the capillary number is low. When we drink a glass of water, for instance, the velocity at which the glass is emptied is of the order of one cm/s, and the capillary number is 10^{-4}. The LLD model is then suitable for evaluating the thickness of the residual film. The result is about 6 μm. When you scramble out of your bath, perhaps to answer an untimely phone call, V is of the order of 1 m/s (the validity of the LLD model is then borderline), in which case you are covered by a thickness of about 150 μm of water. On an average adult human body (about 1.7 m²), this corresponds to a weight of about 250 g.

When the capillary number approaches 1, both the thickness and the extent of the dynamical meniscus tend toward the capillary length. Derjaguin has shown that gravity then becomes the dominant force, and limits the thickness of the film. Up to this point, we have neglected the effect of gravity in comparison to that of surface of tension. The condition for this approximation to hold is $\rho g \ll \gamma\kappa/l$ (gravity negligible when compared to the gradient of the Laplace pressure). Again, this leads to the condition $l \ll \kappa^{-1}$, which is equivalent to the criterion (5.32). When Ca exceeds unity, the LLD visco-capillary regime is replaced by a visco-gravitational one. In this case, viscous and gravitational forces balance each other out, and equation (5.27) must be rewritten as

$$\frac{\eta V}{e^2} \approx \rho g. \tag{5.33}$$

From this we easily derive *Derjaguin's law*:[16]

$$\boxed{e \approx \kappa^{-1} Ca^{1/2}} \tag{5.34}$$

which does indeed merge seamlessly with the LLD law [equation (5.30)] when $Ca = 1$, in which case the thickness of the deposited liquid layer is equal to the capillary length. Equation (5.34) can be derived from an even more direct argument. Since a gravity regime takes the place of a capillary regime, Derjaguin looked for a law of the form $\kappa^{-1} Ca^n$, where the exponent n must be such as to make the latter expression independent of γ.[16] This happens only when $n = 1/2$.

5.3.2 Soapy Liquids

We have assumed up to now that the liquid being dragged is *pure*. That is rarely the case in real life. In most practical situations, the liquids one deals with are dispersed mixtures (suspensions or emulsions) containing surfactants (a detailed discussion of surfactants is deferred until chapter 8). Even in the supposedly "simple" case of soap water, the problem proves far more difficult than in the case of pure water. We will see later on that the main effect of a surfactant is to lower the surface tension of water (typically by a factor of 2). This has repercussions on the capillary length as well as on the capillary number, both of which intervene in the LLD law. But the primary difficulty has to do with the fact that when a solid is being pulled out of a liquid, the surface gets diluted, which creates a concentration gradient of surfactant at the surface, and thus a gradient of surface tension. The physical consequence is a certain "stiffening" of the free surface, which causes it to behave somewhat like a solid. The surface can become the seat of a viscous stress, which can be balanced by the gradient in surface tension.

When a plate is pulled out of soap water, it is as though two interfaces (as opposed to just one as in the case of pure water) drag the liquid along with them, which results in a film about twice as thick as it otherwise would be.[17] The precise value of the extra thickness is difficult to calculate. It has to depend on the surfactant concentration since the behavior of a pure liquid must be recovered at very low concentrations. Nevertheless, it is useful to hold on to the qualitative notion of a soapy interface endowed with *drawing power*, that is to say, with the ability to drag a liquid along with it, very much as a solid would. As it turns out, there is in this respect a very telling experiment, which is the venerable art of soap-bubble making. When one blows a bubble, or when one creates a foam, some matter is being dragged by viscosity *without the presence of any solid "below."* It is the soapy interfaces themselves that effectively play the role of the solid. For low capillary numbers, the film thickness is determined by an LLD type of law, which in this particular case is known as *Frankel's law.*[18] It stipulates that the thickness of a soap bubble at the moment of its birth is 2e [where e is given by equation (5.30)] since two interfaces participate in making the film.

5.3.3   Other Geometries

Up to this point, we have considered liquids deposited on plates or other large objects. What happens in the case of smaller objects, such as tubes or threads?[19] Such questions have direct practical applications. The fact that some residual liquid remains in a tube when one tries to empty it out has important ramifications in the process of assisted recovery of petroleum (when one pumps out a porous rock saturated with oil, approximately 40% of the crude oil is left behind). On a more modest laboratory scale, most of us are familiar with the propensity of a pipette to retain a small amount of liquid after it has been drained out. At the other end of the spectrum, the "greasing" or "oiling" of fibers, which refers to their lubrication at high speed, is an important industrial process. It benefits both their manufacture (it cuts back on ruptures by improving their cohesiveness) and their applications (if the fibers are intended to be used as reinforcement of composite materials, the process is used to coat the fibers with adhesion-promoting substances).

If a change in scale also changes the rate of deposition, it is because the statics are different. We have already discussed the static meniscus that develops inside a tube or on a fiber. In particular, if the radius b characteristic of the object is less than the capillary length (of sub-millimeter size for practical purposes), then the characteristic size of the meniscus is b as well. We have also established that the curvature of the meniscus is constant (since gravity is negligible) and that it is equal to $-2/b$ in a tube and 0 on a fiber.

Consider now the dynamical deposition of a film caused by a relative motion between the solid and the liquid (Figure 5.11). For low capillary numbers, the thickness of the film results from a compromise similar to the one we have described in the context of a plate. Specifically, the viscous force in the meniscus is balanced out by the capillary force which tends to bring the liquid back into the reservoir. Remarkably, in both geometries considered here, the expression giving the Laplace pressure difference between the film and the reservoir takes on the same form when the thickness is small. In this limit, the film turns out to be in state of overpressure by an amount γ/b with respect to the reservoir.

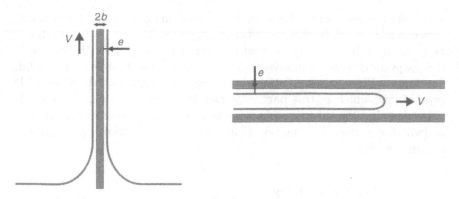

FIGURE 5.11. Fiber being drawn from a liquid pool or capillary tube being drained out. In both cases, a film is deposited on the solid wall.

Since the characteristic thickness of the dynamical meniscus is e, the viscous force takes on the functional form written in equation (5.27) and is in equilibrium with the gradient of the Laplace pressure prevailing within the dynamical meniscus of length l. Therefore

$$\frac{\eta V}{e^2} \approx \frac{\gamma}{lb}. \tag{5.35}$$

In the present configuration, the LLD-like matching condition between the dynamical and static menisci reads

$$\frac{e}{l^2} \approx \frac{1}{b}. \tag{5.36}$$

It is straightforward to deduce from the last two equations the thickness of the deposited layer, known as *Bretherton's law* in the case of a tube:[20]

$$e \approx bCa^{2/3}. \tag{5.37}$$

The fact that a single law covers both geometries considered here makes sense only in the limit $e \ll b$, in other words, for low capillary numbers. When this condition is not met (as depicted in Figure 5.11), geometrical effects come into the picture that introduce differences between fibers and tubes. As soon as the thickness becomes relatively large, the curvature of the film being drawn out changes to $b + e$ for a fiber, and to $b - e$ for a tube. As a result, the quantity b in equation (5.37) must be replaced by either $b + e$ or $b - e$, as applicable. The two dependencies then become quite different. For a fiber, the law becomes $e \approx (b + e)Ca^{2/3}$, which implies a *divergence* of the thickness when Ca approaches unity.[19, 21] On the contrary, the law $e \approx (b - e)Ca^{2/3}$ for a capillary tube implies a *convergence*, which is clearly necessary since in this case $e < b$. Taylor has demonstrated experimentally that the thickness e saturates at about $0.35b$ for large values of Ca.[22]

Finally, we mention the existence of inertial effects in the case of large velocities—an important situation in practice—which complicate the analysis. In a first regime, the thickness increases because of the inertia of the liquid being dragged along with the solid. This effect is observed when the kinetic energy ρV^2 of the liquid (per unit volume) becomes comparable to the Laplace pressure γ/b. The corresponding velocity is of the order of 10 cm/s to 1 m/s for radii between 1 mm and 10 μm. At higher velocities, the thickness decreases because of the inertia of the reservoir standing still. The film is then just the viscous boundary layer that has developed in the reservoir, and that scales as $(\eta L/\rho V)^{1/2}$, where L denotes the reservoir length. All these regimes have been analyzed in detail in the literature.[19, 23]

5.4 Dynamics of Impregnation

5.4.1 Description of the Phenomenon

Impregnation is yet another classic phenomenon in the field of interfacial dynamics. We restrict ourselves to the example of a capillary tube of radius R placed in contact with a reservoir of liquid (Figure 5.12).

As we have seen in chapter 2, invasion of the tube by the liquid proceeds if the surface energy of the wall decreases with wetting ($\gamma_{SL} < \gamma_{SO}$). The force F that causes the liquid to move in the tube is [equation (2.34)]

$$F = 2\pi R\gamma \cos\theta_E \qquad (5.38)$$

where we have used Young's relation to express F in terms of the contact angle θ_E. When the tube is vertical, equilibrium of this force with the weight $W = Mg$ of the liquid column gives the ultimate height (Jurin's law).

From the point of view of dynamics, two forces oppose the movement of the fluid. They are its inertia and the force F_η of viscous friction, respectively. Thus, the equation of motion can be written as

$$\frac{d(MV)}{dt} = F - F_\eta - W \qquad (5.39)$$

where V denotes the average velocity of the fluid.

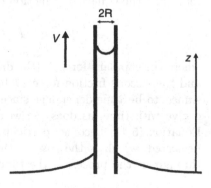

FIGURE 5.12. Capillary rise. The meniscus has a height z (less than Jurin's height) and progresses at a velocity V.

The viscous friction force in a tube is given by Poiseuille's law. The velocity gradients in the tube are of order V/R, which means that the viscous stress is of order $\eta V/R$. In a tube invaded over a length z (hence over a surface $2\pi Rz$), we conclude that the force increases as ηVz. The numerical coefficient in this law is 8π.[16] Therefore, the force F_η can be written as

$$F_\eta = 8\pi\eta Vz. \tag{5.40}$$

Note that equation (5.40) neglects the viscous dissipation associated with the displacement of the contact line—a topic to be discussed in chapter 6. We will see then that this simplification assumes z to be rather large ($z \geq 10R/\theta_e$).

> **Problem:** By writing the Stokes equation (Navier-Stokes equation without the inertial term) in cylindrical coordinates, show that the velocity profile in the tube is parabolic.[24] Designate by G the pressure gradient responsible for the transport of fluid (in the present problem, G is generated by the Laplace underpressure existing at the upstream interface), and set the velocity at the solid/liquid interface equal to zero. Under these conditions, deduce the average velocity in the tube (Poiseuille's law) and, from there, the viscous force F_η that opposes the progress of the fluid [equation (5.40)].

We proceed now to discuss how equation (5.40) simplifies in two different limiting cases, which will enable us to gain further insight into how the liquid progresses in the tube.

5.4.2 Washburn's Law

Tubes (or, more generally, pores) through which liquids are made to flow in impregnation experiments are usually small enough for the movement of fluids to be slow. After all, in a small tube, all molecules are relatively close to a wall where the velocity must vanish. Under these conditions, the inertia of the fluid can be neglected in comparison to the viscous friction term, and the equation of motion reads

$$F - W = F_\eta \tag{5.41}$$

where the capillary force F (the driving term) is given by equation (5.38), and the viscous friction force F_η by equation (5.40). Note that F_η is presumed to be time-dependent since both the height z and the velocity V evolve with time, as does the weight of the liquid column $W = \rho g\pi R^2 z$. Equation (5.41) becomes particularly simple when the weight W can be neglected, which is the case in the early stages of the capillary rise, or throughout the process if the tube is horizontal. With these restrictions, equation (5.41) simplifies to $F = F_\eta$, whose solution is

$$z^2(t) = \frac{1}{2}\frac{\gamma R \cos\theta_E}{\eta}t. \tag{5.42}$$

This result is often referred to as *Washburn's law*.[25] It describes well the experimental finding that the progress of the liquid starts out fast and subsequently slows down ($z \propto \sqrt{t}$). That is certainly consistent with what we ordinarily see when oil is spilled on a piece of cloth or a drop of coffee lands on sugar. It is an important law because it makes it possible in principle to determine from impregnation experiments the value of the contact angle θ_E—a quantity that is notoriously difficult to measure on a porous material. However, experiments turn out to be more difficult to carry out because of the need to first calibrate them in order to know the radius R characteristic of the porous medium. A more detailed description of wetting in porous media will be given in section 9.3.

Note finally that the weight may quickly cease to be negligible, particular for vertical tubes. Equation (5.41) can be integrated completely to find the detailed dynamical properties of a capillary rise experiment (the dedicated reader might want to try it as an exercise). More simply, one could argue that toward the end of the phenomenon, a good approximation of the friction force is $8\pi\eta V z_0$, where z_0 is Jurin's equilibrium height [equation (2.31)]. In the dynamical equation, the velocity is proportional to the difference ($z_0 - z$), which means that the late stage of the rise is simply an exponential relaxation toward Jurin's height.

5.4.3 Inertial Regime

There is a natural limitation to Washburn's law for very short times ($t \to 0$). The problem is that the impregnation velocity (which varies as $1/\sqrt{t}$) then diverges, which is physically impossible. What restores sanity to the situation is the inertia of the liquid, which was neglected in Washburn's model. The tube (or the porous medium) is connected to a vessel containing a liquid at rest, which resists sudden movements. Since this early phase of the process occurs for short times as well as for small heights ($z \to 0$), we may neglect in equation (5.39) both the viscous friction force F_η and the weight W. We are then left with

$$\frac{d(MV)}{dt} = F \tag{5.43}$$

where the driving force F is constant, but where both the mass M of the liquid in motion* and the velocity V are a priori time-dependent. Although

---

*Recall that $M(t) = \rho\pi R^2 z(t)$.

this equation is non-linear, it happens to have a remarkably simple solution corresponding to a constant velocity:[26]

$$z(t) = \left(\frac{2\gamma \cos \theta_E}{\rho R}\right)^{1/2} t. \qquad (5.44)$$

The impregnation velocity defined by this equation is about 30 cm/s for a millimeter-size tube.

We conclude this analysis with three remarks:

1. The constant velocity solution ultimately catches up to Washburn's diffusive solution. This defines a time τ below which inertia prevails over viscosity.[27] This time is given dimensionally by $\rho R^2/\eta$ and can approach 1 s for a millimeter-size tube and for water. The quantity τ can also be interpreted as the time required for a viscous flow to establish itself in the tube. This happens as soon as the liquid enters the tube by way of diffusion of the viscous boundary layer (of thickness δ). The progress of this layer obeys Prandtl's law ($\delta \approx \sqrt{\eta t/\rho}$). It therefore requires a time τ for the layer to make its way from the wall to the center of the tube ($\delta \approx R$).

2. It is instructive to discuss the establishment of the inertial regime itself. The liquid at rest in the vessel acquires a constant velocity defined by equation (5.44) in a time less than τ. This transient regime is quite short, and it is not overly important to characterize it in practice. The only interesting case is that of τ when it is very long, which implies that R is large. While here on earth, as we have seen, R is always limited by the capillary length (beyond which gravity dominates), the same does not hold true in space. In the absence of gravity, capillarity is an interesting alternative for moving liquids around. In order to feed satellite motors with fuel, for instance, one can use a set of parallel plates separated typically by a centimeter to connect the tank to the motor. Upon opening a shutter, the liquid makes its way between the plates by capillarity. In this case, inertial regimes dominate (τ can be of the order of a minute), and the way the fluid is set into motion can be observed.[28]

3. It is always intriguing to discover clouds of uncertainties where everything seems at first to fit together logically. In this spirit, we may take a fresh look at the issue of the inertial regime from an energy perspective. Since in this regime we neglect both the weight of the liquid and viscous friction, this form of impregnation can be viewed as a conversion of surface energy into kinetic energy. The energy per unit area gained as the liquid invades the tube is ($\gamma_{SO} - \gamma_{SL}$) (which, by virtue of Young's relation, is equal to $\gamma \cos \theta_E$), whereas the kinetic energy is $mV^2/2$, where m designates the mass of liquid per unit area. From there we can indeed deduce a constant impregnation velocity which has exactly the same dimensional form as equation (5.44), but with a numerical coefficient of 4 instead of 2. This discrepancy implies some form of energy loss incurred

by the liquid as it enters the tube. This energy loss may be associated with vortices forming at the entrance of the tube on the reservoir side.[29]

5.5   Waves and Ripples

5.5.1   Deep Water Condition

In section 5.2, we have discussed the types of waves that can form —often spontaneously as in the case of the Plateau-Rayleigh or Rayleigh-Taylor instabilities—at the surface of a thin liquid film. Another classic case is that of waves developing in deep waters under the influence of an external perturbation (such as a stone tossed in a pond).[24, 30–32] We are interested here in the dynamics of such disturbances, which we will call either waves or ripples depending on whether they are dominated by gravity or capillarity. As we have seen earlier, another way to distinguish the two cases is in terms of the wavelength (the wavelength of waves will be greater than the capillary length and that of ripples will be smaller). Let ζ be the amplitude of the waves and q their wavevector ($\lambda = 2\pi/q$).

The equation describing the propagation of these waves can be written in the form of equation (5.1) after dropping the viscous term (which is most often negligible in a case involving deep waters):

$$\rho\frac{\partial v}{\partial t} = -\nabla p + \rho g \tag{5.45}$$

where the spatial derivatives $(v \cdot \nabla)v$ have been neglected in writing dv/dt. During a characteristic oscillation time t, an element of fluid covers a distance of order ζ (the amplitude of the wave), which implies a velocity $v \approx \zeta/t$. This velocity is subject to time variations over a time t and spatial variations over a distance of the order of the wavelength λ of the oscillation. The condition $(v \cdot \nabla)v \ll \partial v/\partial t$ can thus be written dimensionally as $v^2/\zeta \ll v/t$. This is equivalent to the recurring small amplitude condition $\zeta \ll \lambda$.

We take as the origin of altitudes z the free surface at rest ($z = 0$). The liquid occupies the region $z < 0$. During motion, there will be a small height deviation ζ. The vertical velocity at the surface is $v_z|_{z=0} = \partial\zeta/\partial t$. In equation (5.45), we resolve the pressure into three separate terms:

$$p = p_0 - \rho g z + \tilde{p} \tag{5.46}$$

where p_0 is the atmospheric pressure. The first two terms represent the static pressure. The third term $\tilde{p}$ represents the pressure modulation. We look for sinusoidal modes of the form

$$\tilde{p}(x, z, t) = \tilde{p}(z) \cdot \exp[i(\omega t + qx)]$$

$$\vec{v}(x, z, t) = \vec{v}(z) \cdot \exp[i(\omega t + qx)] \tag{5.47}$$

$$\zeta(x, t) = \zeta \cdot \exp[i(\omega t + qx)].$$

Equation (5.45) indicates that the velocity $\vec{v}$ is a gradient:

$$\vec{v} = -\frac{1}{i\omega\rho}\vec{\nabla}\tilde{p} \qquad (5.48)$$

whose flow is irrotational. Furthermore, fluids are practically incompressible, which is expressed as $\nabla \cdot \vec{v} = 0$. If so, equation (5.48) then implies:

$$\nabla^2\tilde{p} = 0$$

and

$$\frac{\partial^2\tilde{p}}{\partial z^2} - q^2\tilde{p}(z) = 0. \qquad (5.49)$$

It follows that $\tilde{p}$ is proportional to $\exp(\pm qz)$. To avoid any divergence at large depths, we must choose

$$\tilde{p}(z) = \tilde{p}_S \exp(qz). \qquad (5.50)$$

Immediately underneath the surface, the local pressure must equal the Laplace pressure:

$$p = p_0 - \gamma\frac{\partial^2\zeta}{\partial x^2} = p_0 + \gamma q^2\zeta. \qquad (5.51)$$

Therefore, in accordance with equation (5.46), the pressure $\tilde{p}_S$ is

$$\tilde{p}_S = \gamma q^2\zeta + \rho g\zeta = \gamma(q^2 + \kappa^2)\zeta. \qquad (5.52)$$

We may calculate v_z from equations (5.48) and (5.52):

$$v_z|_{z=0} = -\frac{1}{i\omega\rho}\frac{\partial\tilde{p}}{\partial z} = -\frac{q}{i\omega\rho}\tilde{p}_S = -\frac{\gamma q(q^2 + \kappa^2)}{i\omega\rho}\zeta. \qquad (5.53)$$

Since we have $v_z|_{z=0} = i\omega\zeta$, we end up with the following dispersion relation:

$$\omega^2 = \frac{\gamma q}{\rho}(q^2 + \kappa^2) = gq(1 + \kappa^{-2}q^2). \qquad (5.54)$$

5.5.2 Dispersion Relation in the Inertial Regime

As we emphasized earlier, small wavelengths ($\lambda \ll 2\pi\kappa^{-1}$) are dominated by capillarity (ripples), whereas longer wavelengths ($\lambda \gg 2\pi\kappa^{-1}$) are dominated by gravity (waves). The length $2\pi\kappa^{-1}$ is typically of the order of a centimeter.

Note also that a small drop vibrating in air has its wavelength necessarily fixed· by its radius R. Since we are then in the capillary limit ($q \gg \kappa$), equation (5.54) enables us to deduce the dimensional form of the period τ of oscillation. The result is $\tau \propto (\rho R^3/\gamma)^{1/2}$. For a millimeter-size drop, this period is of the order of a millisecond.

Equation (5.54) provides the phase velocity $c = \omega/q$ of the surface waves. We find

$$c = \left(\frac{g}{q} + \frac{\gamma q}{\rho}\right)^{1/2}. \tag{5.55}$$

A surface excited by a tuning fork vibrating at a frequency $\omega/2\pi = 440$ Hz generates capillary waves with a wavelength of the order of a millimeter [equation (5.54)] traveling at a velocity of about 60 cm/s [equation (5.55)]. A remarkable property predicted by equation (5.55) is the existence of a minimum for c. The minimum c^* is obtained when $\lambda = \sqrt{2}(g\lambda/\rho)^{1/4}$. The result is 23 cm/s for water.

A fisherman standing on the edge of a fast-moving river can see around and downstream of his line a remarkable capillary wake that has fascinated physicists (such as Lord Rayleigh, Lamb, Lighthill, and others) for a century.[30, 31] The wake forms only when the speed of the river relative to the line exceeds the minimum velocity c^* of the associated waves. Generally speaking, the entire structure of the wake can be analyzed in detail with the help of equation (5.55). The interested reader may want to catch up with recently published developments on this topic.[33]

All these results must of course be modified if the fluid has a thickness h of the order of or less than the wavelength λ. An example is ocean waves washing up on a beach. If the effect of viscosity can still be neglected, it is necessary to introduce an additional boundary condition, namely, that the vertical component of the velocity v_z be zero at $z = -h$. The result when qh is small is

$$\omega^2 = ghq^2(1 + \kappa^{-2}q^2). \tag{5.56}$$

The phase velocity in the same limit reads

$$c = \sqrt{gh}(1 + \kappa^{-2}q^2)^{1/2} \tag{5.57}$$

which is smaller by a factor of $\sqrt{qh}$ than the result predicted by equation (5.55). Consider, for example, a wave whose period is 10 s. The wavelength and velocity in deep waters are given by equations (5.54) and (5.55) respectively, with numerical values of 100 m and 10 m/s. These two quantities are noticeably reduced upon approaching the shoreline. For a depth of 1 m, the results are 25 m and 2.5 m/s, while for a depth of 15 cm, note the results drop further to 10 m and 1 m/s (these considerations are not without profound ramifications for surfing afficionados). As a final comment, we could refine the analysis to the next level of complication by noting that in the limit of small depths, the crests and troughs of waves no longer have the same velocity, a fact that is responsible for rolling waves.

5.5.3 Attenuation

The arguments we have just used cease to hold in the limit of very small thicknesses (low Reynolds numbers), as discussed in section 2.2. In that case, the primary source of viscous dissipation is the movements of the fluid within the wave itself. We can obtain an order of magnitude of the characteristic time τ for attenuation by writing the equilibrium between inertia and viscous forces in the wave. When doing so, we find $\rho V/\tau \approx \eta V/\lambda^2$. This leads to a characteristic time $\tau \approx \lambda^2/\eta$, which can be quite small for capillary waves (small λ). We can associate with this time a length $\mu = c\tau$, which is the length over which capillary waves of wavelength λ transported at a velocity c [given by equation (5.55)] are attenuated. The result is

$$\mu \approx \frac{(\rho\gamma\lambda^3)^{1/2}}{\eta}. \tag{5.58}$$

For the capillary wave considered above ($\lambda \sim 1$ mm on water), we find that the wave gets attenuated over a distance of 30 cm. This distance decreases rapidly as the wavelength diminishes (it is only 1 cm for $\lambda \sim 100$ μm). We can observe these two lengths λ and μ quite plainly when we aim a small stream of water (such as is produced by a faucet) onto an obstacle. We can then see a stationary pattern of capillary waves at the bottom of the stream [whose wavelength is determined by equation (5.55) after imposing $c = V$ (the velocity of the stream)] making its way back up the stream over a length μ (typically a centimeter). Note that in the case of gravity-driven waves ($\lambda > 2\pi\kappa^{-1}$), the length μ, given then dimensionally by $\rho g^{1/2}\lambda^{5/2}/\eta$, becomes downright huge. For $\lambda = 1$ m, the result turns out to be several kilometers!

References

[1] H. Jeffreys, *Proc. Camb. Phil. Soc.* **26**, 204 (1930).

[2] S. Chandrasekhar, *Hydrodynamic and Hydromagnetic Stability* (Oxford, U.K.: Clarendon Press, 1961).

[3] J. W. Swan, *Proc. Roy. Soc. London* **62**, 38 (1897).

[4] E. Schäfer, T. Thurn–Albrecht, T. P. Russell, and U. Steiner, *Europhys. Lett.* **35**, 518 (2001).

[5] M. D. Cowley, and R. E. Rosensweig, *J. Fluid Mech.* **30**, 671 (1967).

[6] J. Browaeys, J. C. Bacri, C. Flament, S. Neveu, and R. Perzynski, *Europhys. J. B* **9**, 335 (1999).

[7] G. I. Taylor, *Proc. Roy. Soc. London* **201**, 192 (1950).

[8] M. Fermigier, L. Limat, J. E. Wesfreid, P. Boudinet, and C. Quilliet, *J. Fluid Mech.* **236**, 349 (1992).

[9] J. Plateau, *Statique expérimentale et théorique des liquides soumis aux seules forces moéculaires* (Experimental and Theoretical Steady State of Liquids Subjected to Nothing but Molecular Forces) (Paris: Gauthiers-Villars, 1873).

[10] Lord Rayleigh, *Phil. Mag.* **34**, 145 (1892).

[11] F. Brochard, *J. Chem. Phys.* **84**, 4664 (1986).

[12] F. Savart, *Annales de Chimie* **53**, 337 (1833).

[13] S. Tomotika, *Proc. Roy. Soc. London* **153**, 322 (1935).

[14] L. Landau and B. Levich, *Acta Physicochem. USSR* **17**, 42 (1942).

[15] B. V. Derjaguin, *Acta Physicochem. USSR* **20**, 349 (1943).

[16] B. V. Derjaguin and S. M. Levi, *Film Coating Theory* (London: The Focal Press, 1943).

[17] D. Quéré, A. de Ryck, and O. Ou Ramdane, *Europhys. Lett.* **37**, 305 (1997).

[18] K. J. Mysels, K. Shinoda, and S, Frankel, *Soap Films* (London: Pergamon Press, 1959).

[19] D. Quéré and A. de Ryck, *Annales de Physique* **23**, 1 (1998); D. Quéré, *Ann. Rev. Fluid Mech.* **31**, 347 (1999).

[20] F. P. Bretherton, *J. Fluid Mech.* **10**, 166 (1961).

[21] D. A. White and J. A. Talmadge, *A. I. Ch. E. Journal* **12**, 333 (1966).

[22] G. I. Taylor, *J. Fluid Mech.* 10, 161 (1961).

[23] P. Aussillous and D. Quéré, *Phys. Fluids* **12**, 2367 (2000).

[24] L. D. Landau and E. M. Lifshitz, *Fluid Mechanics* (London: Pergamon Press, 1959).

[25] E. W. Washburn, *Phys. Rev.* **17**, 273 (1921).

[26] D. Quéré, *Europhys. Lett.* **39**, 533 (1997).

[27] C. H. Bosanquet, *Phil. Mag.* **45**, 525 (1923).

[28] M. Dreyer, E. Delgado, and H. Rath, *J. Colloid Interface Sci.* **163**, 158 (1994).

[29] D. Quéré, E. Raphaël, and J. Y. Ollitrault, *Langmuir* **15**, 3679 (1999); E. Lorenceau, D. Quéré, J. Y. Ollitrault, and C. Clanet, *Phys. Fluids* **14**, 1985 (2002).

[30] H. Lamb, *Hydrodynamics* (Cambridge, U.K.: Cambridge University Press, 1932).

31 J. Lighthill, *Waves in Fluids* (Cambridge, U.K.: Cambridge University Press, 1978).

32 E. Guyon, J. P. Hulin, and L. Petit, *Hydrodynamique physique* (Physical Hydrodynamics) (Paris: InterEditions, 1991).

33 E. Raphaël and P. G. de Gennes, *Phys. Rev. E* **53**, 3448 (1996).

6

Dynamics of the Triple Line

6.1 Basic Experiment

Consider a *clean* surface, free of hysteresis. We have seen that the contact angle θ_E results from an equilibrium between several forces. When the liquid wedge is set in motion, the dynamical contact angle θ_D is different from θ_E at equilibrium. The force of traction pulling the liquid toward the dry region (Figure 6.1) is given by

$$F(\theta_D) = \gamma_{SO} - \gamma_{SL} - \gamma \cos \theta_D. \tag{6.1}$$

The force vanishes at equilibrium, $F(\theta_E) = 0$. Our purpose here is to consider the situation when the force F is not zero and when the triple line moves with a velocity V.

Most often, the relevant experiment consists in imposing a velocity V and in determining the angle θ_D by optical means (either by photography or, for better precision, by interferometry).

- When $F > 0$, the liquid contained in a capillary tube can be forced to move by pushing it with a piston (Figure 6.2).

  The first such measurements were carried out by Hoffmann.[1] Some of his results are reproduced in Figure 6.3 in the case of total wetting ($\theta_E = 0$).

- For $F < 0$, one oftentimes resorts to "extraction" experiments in which a vertical plate is pulled out vertically at a velocity V_p from a liquid bath (Figure 6.4). Under such conditions, a stationary regime can be set up where the triple line remains at a fixed height. In other words, it then moves at a velocity $V = -V_p$ relative to the plate (Figure 6.4a).

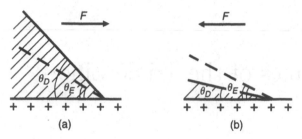

FIGURE 6.1. Non-equilibrium force exerted on the triple line: (a) $F > 0$; (b) $F < 0$.

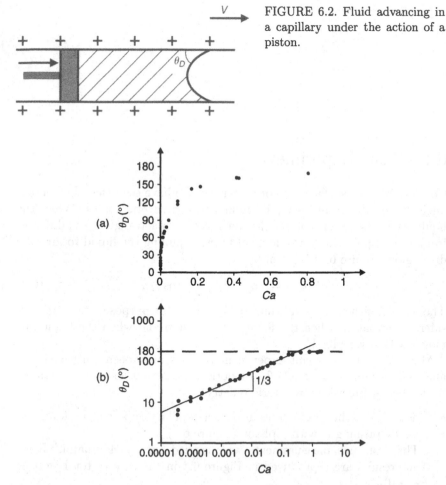

FIGURE 6.2. Fluid advancing in a capillary under the action of a piston.

FIGURE 6.3. Dynamical contact angle θ_D as a function of the capillary number $Ca = \eta V/\gamma$ ($\eta =$ viscosity, $\gamma =$ surface tension, $V =$ velocity of the triple line) for various silicone oils in a glass tube (data from ref. 1). This particular case corresponds to total wetting ($\theta_E = 0$). (a) Linear scale and (b) log scale, showing the $V \propto \theta_D^3$ dependence.

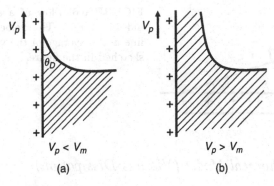

$$V_p < V_m \qquad\qquad V_p > V_m$$

(a)   (b)

FIGURE 6.4. Vertical extraction of a plate from a pool of liquid.

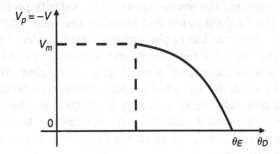

FIGURE 6.5. Vertical extraction: Dependence of the velocity on the dynamical angle θ_D.

However, this regime can be observed only if the pulling rate is not too high. Beyond a certain threshold velocity V_m, the triple line no longer finds a stable position, and the plate drags along with it a film of finite thickness (Figure 6.4b). This phenomenon has been discussed in chapter 5 in the context of forced wetting. The relationship between the angle θ_D and the velocity V is shown in Figure 6.5. Our next objective is to understand the nature of $F(V)$.

6.2   Relation Between Force and Velocity

The dynamical properties of the triple line involve local phenomena—on a molecular scale—in the immediate vicinity of the line, as well as longer-range phenomena in the form of viscous flows in the overall liquid body. We shall see that both aspects are generally at play. Only in one particular case (when θ is small) do macroscopic effects tend to dominate.

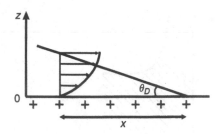

FIGURE 6.6. Flow of a spreading liquid ($\theta_D > \theta_E$). The velocity V of the line is the average of the velocity profile sketched in the figure.

6.2.1 Mechanical Model (Viscous Dissipation)

We focus our attention on the case when $\theta \ll 1$, which is simpler to treat and more important conceptually. We consider a perfect wedge-shaped liquid border characterized by an angle θ, as depicted in Figure 6.6. The slope $\tan\theta \approx \theta$ is a constant. The wedge moves with a velocity profile $v(z)$ starting at 0 near the solid substrate and reaching the value $1.5V$ at the top. Note that the velocity at the upper surface is greater than the advancing velocity V of the line (V is the velocity averaged over the thickness of the wedge). This was demonstrated in an elegant experiment done by Elizabeth Dussan, in which she deposited a colored marker on the upper surface. If you were to duplicate the experiment, you would see the marker catch up to the line and fix itself onto the solid substrate. The stain advances somewhat like a tank's tread, which moves faster than the tank itself when not in contact with the ground.

Returning to the problem at hand, the velocity varies with the height z, from 0 at $z = 0$ to $1.5V$ at $z = \theta_D x$. There exists, therefore, a velocity gradient:

$$\frac{dv}{dz} \approx \frac{V}{\theta_D x}.$$

We next calculate the entropy source $\dot{S}$ or the energy $T\dot{S}$ dissipated by viscous phenomena (expressed per unit length of the line in the y-direction). Using η to designate the viscosity, we start with the general expression

$$T\dot{S} = \int_0^\infty dx \int_0^e \eta \left(\frac{dv}{dz}\right)^2 dz. \qquad (6.2)$$

In the present case, this becomes

$$T\dot{S} \cong \int_0^\infty dx \cdot \eta \frac{V^2}{e^2} e. \qquad (6.3)$$

Thus, we get

$$T\dot{S} = \frac{3\eta V^2}{\theta_D} \int_0^\infty \frac{dx}{x}. \qquad (6.4)$$

The exact prefactor has been restored in the last equation. The integral in equation (6.4) diverges at both limits. We can take care of this problem by truncating the integral on the high side at $x = L$ (the size of the drop) and on the low side at $x = a$ (the molecular size). Within these constraints, we have *

$$\int_a^L \frac{dx}{x} = \ln\left(\frac{L}{a}\right) \equiv l. \tag{6.5}$$

The numerical value of the dimensionless coefficient l ranges from 15 to 20. Readers should refer to the published literature for a more extensive discussion of the coefficient l.[2,3]

We are now in a position to derive the dependence of $F(V)$ by equating two forms of the dissipation:

$$T\dot{S} = FV = \frac{3\eta l}{\theta_D}V^2. \tag{6.6}$$

Returning to the definition of the force [equation (6.1)] and in the small angle approximation ($\cos\theta \approx 1 - \theta^2/2$), we have

$$V = \frac{V^*}{6l}\theta_D \cdot (\theta_D^2 - \theta_E^2). \tag{6.7}$$

The velocity $V^* \equiv \dot{\gamma}/\eta$ is a recurring parameter coming up in most capillary phenomena in viscous regimes. Its value is typically 30 m/s.

Hoffmann's experimental data are reproduced in Figure 6.3.[1] They correspond to total wetting ($\theta_E = 0$) and are entirely consistent with a θ_D^3 dependence, as expressed in equation (6.7).

The functional dependence of $V(\theta_D)$ is quite remarkable. When $\theta_D \to \theta_E$, one finds $V = 0$, which corresponds to static equilibrium. However, one also finds $V \to 0$ when $\theta_D \to 0$. This means that the dissipation in a sharply angled wedge is extremely strong and impedes any motion. Between these two points is a minimum of V, i.e., a maximum of $V_p = -V$, which we denote V_m. In the experiment illustrated in Figure 6.4, there can be no meniscus at extraction velocities exceeding V_m. As the extraction velocity increases and goes through V_m, a sudden transition occurs from a meniscus regime to a film regime.

From equation (6.7) we derive

$$V_m = \frac{V^*}{9\sqrt{3}l}\theta_E^3.$$

In water, $V^* = 70$ m/s. With $\theta_E = 0.1$ radian and $l = 20$, the result is $V_m = 0.2$ mm/s. Such velocities have a bearing on the fabrication of so-called Langmuir-Blodgett layers, when a film of surfactant is being transferred from the free surface of a liquid to a plate being drawn out vertically. Such a transfer is possible only if the free surface comes into contact with the plate, that is to say, when a meniscus is present. This requires that the extraction speed be less than V_m.

---

*Strictly speaking, the coefficient l can depend logarithmically on the velocity if we use a/θ_D as the lower bound of the integral 6.5. A weak dependence of l on V has indeed been observed by Marsh et al. (see ref. 2).

Note: Correction to an Ideal Wedge

Laplace's equation (1.6) states that a profile with a constant slope (i.e., with zero curvature) requires the pressure to be equal to the atmospheric pressure p_0 throughout the liquid. This approximation cannot be strictly correct because the flow induces pressure gradients. The profile is determined by Poiseuille's equation:

$$-\frac{\partial p}{\partial x} = \eta \frac{\partial^2 v}{\partial z^2} \approx \eta \frac{V}{e^2} \tag{6.8}$$

which implies a pressure given by

$$p(x) - p_0 \approx \left(\int_x^\infty -\frac{\eta V}{e^2} \right) dx. \tag{6.9}$$

The rationale for the last equation is that the pressure must merge with the atmospheric pressure for large values of x. Since $e = \theta_D x$, we get

$$p - p_0 \approx -\frac{\eta V}{\theta_D^2 x}. \tag{6.10}$$

Laplace's law therefore imposes a slight curvature given by

$$p - p_0 = -\gamma \frac{\partial^2 e}{\partial x^2}. \tag{6.11}$$

Integrating once gives a correction to the slope:

$$\left. \frac{\partial e}{\partial x} \right|_1 = \frac{V}{V^*} \frac{1}{\theta_D^2} \ln\left(\frac{L}{x}\right). \tag{6.12}$$

This logarithmic correction has been found experimentally.[4] In order for the preceding analysis to be valid, it is necessary that the correction be small in comparison with θ_D. Omitting the logarithmic factors, the relative correction is

$$\frac{1}{\theta_D} \frac{\partial e_1}{\partial x} \approx \frac{V}{V^*} \frac{1}{\theta_D^3} \approx \frac{\theta_D^2 - \theta_E^2}{\theta_D^2}. \tag{6.13}$$

It becomes clear that the approximation is a good one as long as $|\theta_D - \theta_E| \ll \theta_E$.

6.2.2 Chemical Model

An alternate point of view is to treat the wetting process as a chemical reaction in which a molecule of the liquid approaches from the upper side of the wedge and binds on the solid substrate.[5] If the local contact angle—on a molecular scale—retains the value θ_D, the force involved (per unit length) remains F as defined before. The displacement during a jump toward the solid takes place over a molecular length a, and the energy liberated during the reaction is Fa^2. However, there can exist an energy barrier U standing in the way of the jump. This leads to a jump frequency (defined as the difference between an adsorption term and a desorption term):

$$\frac{1}{\tau} = \frac{1}{\tau_0}\left\{\exp\left(\frac{-U + Fa^2/2}{kT}\right) - \exp\left(\frac{-U - Fa^2/2}{kT}\right)\right\} \qquad (6.14)$$

where τ_0 is a microscopic time of the order of 10^{-11} s. We can write the velocity as $V = a/\tau$. In the linear regime (when the forces are weak), we obtain a functional relationship between force and velocity of the form:

$$V = V_0 \exp\left(-\frac{U}{kT}\right)\frac{Fa^2}{kT}. \qquad (6.15)$$

This implies a dissipation given by

$$T\dot{S} = FV = \eta_1 V^2 \qquad (6.16)$$

where the coefficient η_1 is defined as

$$\eta_1 \approx \frac{kT}{V_0 a^2}\exp\left(+\frac{U}{kT}\right). \qquad (6.17)$$

Equation (6.16) resembles equation (6.6), with two important differences:

1. The viscosity η of the liquid depends strongly on temperature and has a certain activation energy U_η, whereas the equivalent viscosity η_1 in the chemical model has a different activation energy U. It is tempting to argue that the reaction might correspond to a molecule jumping from the liquid to the wall by way of a brief transition through a vapor phase. If so, U should be comparable to the vaporization heat (per molecule) L_v.[6] This interpretation leads to $\exp(-U/kT) \approx \rho_{gas}/\rho_{liq} \approx 10^{-3}$, where the quantities ρ are densities. The most reasonable approach to understand fully these complicated effects is to simulate a small drop (typically 2,000 atoms) in the process of spreading.[7]
2. The "jump length" a which has to be inserted in the equations to account for the experimental data turns out to be abnormally large (orders of magnitude larger than the molecular size).

Is there a way to declare a winner between the hydrodynamic model and the chemical one? Based on the data of Figure 6.3, the angle θ_D depends only on the product ηV. This observation suggests that the viscosity η controls the friction, in which case the hydrodynamic model prevails. Moreover, we see that $V \propto \theta_D^3$ when $\theta_D < 1$ in the total wetting regime, precisely as the hydrodynamic model predicts.

In the macroscopic model, the dissipation is inversely proportional to θ_D. This property has no obvious counterpart in the chemical model. The fact that the dissipation becomes large when $\theta_D \to 0$ is responsible for the existence of a maximum velocity V_m at a finite value of θ_D (when $\theta_D = \theta_m$). This suggests the need to measure carefully the angle θ_m at which the meniscus/film transition takes place during withdrawal experiments. When

$\theta_m = 0$, "chemical" effects predominate. When $\theta_m > 0$, hydrodynamic effects become increasingly important. As of this writing, there have been no systematic experimental studies to resolve the issue.

Conclusion. The chemical model is somewhat fuzzy in its details. Just what is the molecular process involved? Is the local contact angle the same as the macroscopic one? Notwithstanding these difficulties, it is clear that microscopic mechanisms in the immediate vicinity of the line can be important for large angles θ_D. A "hybrid" model, in which the hydrodynamic and molecular dissipations are simply added together, has been proposed.[8] Nevertheless, the assumption that the two contributions are additive is far from being obvious.

6.3   Oscillations Modes of a Triple Line

A triple line can vibrate somewhat in the manner of a guitar string. Unlike the case of a string, however, two effects must be included here:

1. The relevant elasticity is of the fringe type (see section 3.2.2).
2. On a solid substrate, and particularly for small angles θ_E, viscous friction is important. Indeed, we will specifically focus on a viscous regime (in the particular case when $\theta_E \ll 1$) as discussed elsewhere.[9]

The fringe elasticity has been discussed in chapter 3 (where we chose $\theta_E = \pi/2$). Here, when the angle θ_D is small, and for a mode of wavevector q and amplitude u_q, the energy retains the same functional dependence on q but with a different prefactor:

$$E = \frac{1}{2}\gamma\theta_E^2 |q| u_q^2. \qquad (6.18)$$

The force is given by

$$-\frac{\partial E}{\partial u_q} = -\gamma\theta_E^2 |q| u_q. \qquad (6.19)$$

A viscous force can balance this force out, which we write as

$$-\gamma\theta_E^2 |q| u_q = -\frac{3\eta l}{\theta_E}\frac{du_q}{dt}. \qquad (6.20)$$

This gives an exponential relaxation, with a relaxation frequency:

$$\frac{1}{\tau_q} = \frac{V^*\theta_E^3}{3l}|q| \equiv c|q|. \qquad (6.21)$$

The dynamics is characterized by a velocity $c \cong V^*\theta_E^3$. The physical meaning of this velocity is illustrated in Figure 6.7.

The modes for a given q have been studied by Ondarçuhu.[10]

FIGURE 6.7. Dynamics of the relaxation of a triple line. At first, the line is pinched by a localized defect and takes on its equilibrium shape, as discussed in section 3.2.2. As the defect is eliminated at time $t = 0$, the line relaxes and reverts to a horizontal shape. Only a region of size ct is relaxed at time t.

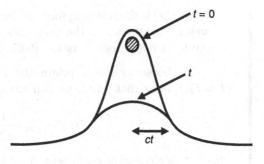

FIGURE 6.8. Beginning of the relaxation process $(t > 0)$. The topmost point of the line has moved down by a distance x.

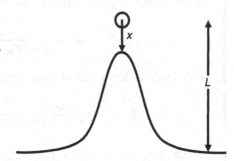

Problem: Describe the "snapping" process of a triple line at a defect.[11]

Answer: The situation of interest is sketched in Figure 6.8. If the defect is strong enough, it is possible to see the line stretch over a large distance $\Delta x \cong L$ (where L is the separation between defects) before snapping.

a) The *return time* τ_r to a straight line is approximately the time τ_q figuring in equation (6.21) with $q \approx 1/L$:

$$\tau_r \cong L/c. \tag{6.22}$$

b) In a static experiment, however, another long-time constant can appear when one increases the macroscopic force F [proportional to $\gamma(\cos\theta_D - \cos\theta_E)$] applied to the line very slowly.* Eventually, a threshold F^* for release is reached, at which point the initial velocity is zero. Stated differently, the time constant τ_i for the initiation of snapping is long. This time constant can be estimated as follows. Let x be the most elongated position along the line. Its motion can be described by

$$\zeta\frac{dx}{dt} = F - f(x) \tag{6.23}$$

---

*This can be accomplished, for instance, by inflating the drop via injection of liquid or by tilting the sample.

where $f(x)$ is the resisting force of the defect, which, when at its maximum, corresponds to the hysteresis threshold F_m. The coefficient ζ can be derived from equation (6.6). The result is $\zeta \propto \eta/\theta_E$.

We choose as the x origin the point where $f(x)$ is maximum $(f = F_m)$. Near that point, we can write

$$f(x) = F_m - \frac{1}{2}f''x^2 \tag{6.24}$$

where f'' is a constant coefficient given by $f'' \approx F_m/b^2$, b being the size of the defect. Equation (6.23) can be recast in the form

$$dt = \frac{\zeta dx}{F - F_m + (1/2)f''x^2}. \tag{6.25}$$

This leads to a net initial time constant given by

$$\tau_i = \int_0^\infty \frac{\zeta dx}{F - F_m + (1/2)f''x^2} = \frac{\pi\zeta}{[2(F - F_m)f'']^{1/2}} \tag{6.26}$$

which diverges as $1/(F - F_m)^{1/2}$. When F is slightly larger than F_m, the starting phase of the process is the limiting factor. Under these conditions, the velocity of the macroscopic line motion is $V \approx L/\tau_i$, which varies as $(F - F_m)^{1/2}$.

6.4   Dynamics of Total Wetting

L. Tanner, an engineer retired from the British Air Force, was the first to investigate the spreading of a drop on a totally wettable surface.[12] He recognized that a drop could be considered a plane/spherical lens. By observing it in reflection, he determined its focal length, hence its radius of curvature or, equivalently, its contact angle θ_D (Figure 6.9).

Lellah and Marmur performed a different but equivalent set of measurements.[13] They monitored the increase in the horizontal radius R(t) of a wet spot. The spreading lasts typically from a few hours for ordinary liq-

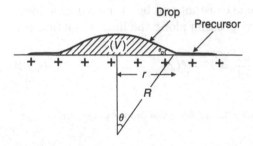

FIGURE 6.9. Spreading of a drop on a solid surface in total wetting regime. With the naked eye, an observer can see the drop, but not the precursor film.

uids to several weeks for highly viscous fluids such as heavy silicone oils. The main practical challenge in these types of experiments is to avoid any contamination (either of the drop or of the surface) over such long times.

The results can be expressed in terms of a contact angle θ_D which depends on the spreading time t. When surfaces are smooth and clean, and for non-volatile liquids, the law is remarkably universal. It is known as Tanner's law:

$$\theta_D(t) \propto t^{-3/10} \qquad (6.27)$$

The measurements reveal a highly surprising fact, namely, that the angle θ_D is completely *independent of the spreading coefficient* S defined in chapter 1 as long as S is positive, that is to say, as long as we are in a total wetting regime. This is surprising because the force F that acts on the system of interest is essentially equal to S:

$$F = \gamma_{SO} - \gamma_{SL} - \gamma \cos\theta_D \cong \gamma_{SO} - \gamma_{SL} - \gamma = S. \qquad (6.28)$$

The explanation for this puzzle is related to another experimental fact, which Hardy observed long ago.[14] It turns out that ahead of the drop is a *precursor film* a few nanometers thick, which extends much farther out than the drop itself. Hardy detected the existence of this film with very clever techniques. In one of them, he used a tiny glass rod resting on the substrate at a finite angle. As long as the surface remained dry, friction held the rod in place. But as soon as the liquid film reached the rod, its lubricating effect loosened the contact and caused the rod to fall. For a detailed review of these phenomena, the reader is invited to consult the literature.[12]

The precursor film is evidence of the great force F acting on its boundary. The liquid is rapidly drawn toward the periphery in the form of a film whose thickness is roughly the pancake thickness defined in chapter 4. But behind the film, where Tanner physically saw the edge of a drop, the forces involved are quite different. Within the drop are the forces of traction $-\gamma_{SL} - \gamma \cos\theta_D$, whereas in the film proper (characterized by a zero angle) are the forces $\gamma_{SL} + \gamma$. The net force acting on the drop is then only

$$\tilde{F} = \gamma - \gamma \cos\theta_D \cong \gamma \frac{\theta_D^2}{2}. \qquad (6.29)$$

We may then predict the velocity through equation (6.6) after replacing F by $\tilde{F}$. This leads to

$$V \equiv \frac{d\mathrm{R}}{dt} = \frac{\theta_D}{3l\eta}\tilde{F} = \frac{V^*}{6l}\theta_D^3. \qquad (6.30)$$

The volume of the drop is conserved (for a macroscopic drop, the volume of the precursor is negligible). The volume is

$$\Omega = \frac{\pi}{4}\mathrm{R}^3\theta_D. \qquad (6.31)$$

By differentiation, we get

$$\frac{3}{R}\frac{dR}{dt} = -\frac{1}{\theta_D}\frac{d\theta_D}{dt}. \tag{6.32}$$

Neglecting the numerical coefficients, we arrive at

$$\frac{1}{\theta_D}\frac{d\theta_D}{dt} = -\frac{V^*}{R}\theta_D^3. \tag{6.33}$$

Equation (6.31) can be recast in the form

$$R = L\theta_D^{-1/3} \tag{6.34}$$

where $L \approx \Omega^{1/3}$ defines the size of the initial drop. Under these circumstances, we have

$$\frac{d\theta_D}{dt} = -\frac{V^*}{L}\theta_D^{13/3} \tag{6.35}$$

which ultimately leads to

$$\theta_D \approx \left(\frac{L}{V^*t}\right)^{3/10} \tag{6.36}$$

This result is in agreement with the observations reported by both Tanner and Marmur. More detailed discussions of the precursor and its merger with the drop can be found elsewhere.[14]

Finally, equation (6.31) enables us to derive the law describing the time evolution of the drop:

$$R \propto L\left(\frac{V^*t}{L}\right)^{1/10}. \tag{6.37}$$

The growth rate described by equation (6.37) is quite slow.

It is important to bear in mind that this equation applies only when R is less than the capillary length κ^{-1}. When $R > \kappa^{-1}$, gravity must be taken into account.[15]

References

[1] R. Hoffmann, *J. Colloid Interface Sci.* **50**, 228 (1975).

[2] J. A. Marsh, S. Garoff, and E. Dussan, *Phys. Rev. Lett.* **70**, 2778 (1993).

[3] P. G. de Gennes, *Rev. Mod. Phys.* **57**, 827 (1985).

[4] C. G. Ngan and E. Dussan, *J. Fluid Mech.* **118**, 27 (1982).

[5] T. Blake and J. M. Haynes, *J. Colloid Interface Sci.* **30**, 421 (1969).

[6] Y. Pomeau, *C. R. Acad. Sci.* (Paris) (to be published).

[7] M. J. de Ruijter, T. Blake, J. de Coninck, and A. Clark, *Langmuir* **13**, 7293 (1987).

[8] M. J. de Ruijter, M. Charlot, M. Voué, and J. de Coninck, *Langmuir* **16**, 2363 (2000).

[9] F. Brochard and P. G. de Gennes, *Langmuir* **7**, 3216 (1991).

[10] T. Ondarçuhu and M. Veyssié, *Nature* **352**, 418 (1991).

[11] E Raphaël and P. G. de Gennes, *J. Chem. Phys.* **90**, 7577 (1989).

[12] L. H. Tanner, *J. Phys. D: Appl. Phys.* **12**, 1473 (1979).

[13] M. Lellah and A. Marmur, *J. Colloid Interface Sci.* **82**, 518 (1981).

[14] W. B. Hardy, *Phil. Mag.* **38**, 49 (1919).

[15] H. E. Huppert, *J. Fluid Mech.* **121**, 43 (1982).

7
Dewetting

Dewetting refers to the spontaneous withdrawal of a liquid film from a hostile surface, for instance, water on a hydrophobic solid.

As we step out of a shower, we generally use a good absorbing towel to dry ourselves and we may grab a hair dryer to restore a fluffy hairdo. But if we pay a bit of attention, we realize that our skin actually dries spontaneously. Areas free of water form and expand. The process involves nucleation and growth of dry regions. That is the method ducks rely on when they emerge from a pond. Their feathers are extremely hydrophobic and dewet almost instantly. In short, there are three ways to restore a dry state:

1. evaporation, which consumes heat;

2. capillary suction (tissue paper, bath towels, and other porous media);

3. dewetting, if the surface is hydrophobic either naturally or through an appropriate surface treatment.

Although "dewetting" is an ubiquitous phenomenon—we can observe it everyday on a windshield, in a glass of water, or while casually taking a bath—the mechanics governing the retreat of liquid films began to be understood only recently.

Dewetting has direct repercussions on many industrial processes. It is useful whenever there is a need to dry a surface quickly, particularly if a coating of dust particles is to be avoided. Conventional drying by evaporation leaves behind a residue of stains. In a dewetting scenario, by contrast, the liquid tows along with it all impurities as it recedes. That is why dishwasher detergents are formulated in such a way as to promote spontaneous dewetting. Likewise, certain fluorinated shampoos facilitate drying by rendering hair more hydrophobic. Another example is taken from the aviation industry. Before an aircraft takes off in wintry weather, ground crews spray it with a liquid that makes the body and wings non-wettable. This treatment prevents the formation of a continuous film of water likely to turn to ice, whose weight would load down the aircraft and compromise its aerodynamic lift.

Dewetting of a liquid film inserted between a soft, deformable material and a solid substrate produces an adhesive bond. When driving on a wet road, brakes are effective only if the film of water between the road and the tires, which plays the role of lubricant, can undergo dewetting in less than a millisecond. Understanding dewetting is important to prevent hydroplaning.

In biology, dewetting controls the dynamics of adhesion on wet substrates (mushroom spores, living cells).

Conversely, some industrial processes aim to spread liquid films, in which case dewetting is undesirable. Such is the case of adhesive tapes, where the glue must coat the surface uniformly without leaving any gaps. The same problem affects certain products developed for agricultural purposes. They are generally water-based, but foliage happens to be very hydrophobic. It is sometimes useful to add special substances to curb dewetting.

The first studies on the topic of dewetting were concerned with the instability of the lachrymal film.[1] When the protein composition of the tears becomes abnormal, the cornea turns hydrophobic and the eye undergoes dewetting, leading to a condition known as dry eye syndrome. Fatty cosmetic creams irritate the eyes because they, too, render the cornea hydrophobic. Those who wear soft contact lenses must take great care to prevent dewetting of the lachrymal film. Otherwise, the lens sticks to the eye, and prying it loose can be an extremely painful experience.

On a fundamental level, the mechanisms controlling the way liquid films recede are somewhat similar to those involved in phase transitions. The film can be metastable and dewet by nucleation and expansion of dry zones. It can also be unstable and dewet spontaneously through amplification of capillary waves. The latter case is referred to as spinodal dewetting.

Dewetting is a special case of film rupture. The main types of film rupture fall into three categories, depicted in Figure 7.1.

a) The liquid film is deposited directly onto a solid or liquid substrate (Figure 7.1a). We will start with ideal substrates (silanized silicon wafers, liquid substrates) before tackling dirty or rough surfaces, and then moving on to porous surfaces on which a liquid recedes also by way of impregnation.

b) The liquid film is wedged between a solid and a soft material (Figure 7.1b). This case will be discussed in chapter 9.

c) Finally, the liquid film can simply be suspended (Figure 7.1c). This case applies to ultra viscous films (e.g., polymers or molten glass) suspended in air or floating on a highly fluid substrate.

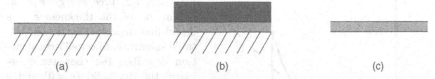

(a)   (b)   (c)

FIGURE 7.1. The dewetting process has been studied in three different geometries: (a) *supported films*, resting on a solid substrate. The process controls the spontaneous drying out of liquid films without the intervention of heat, as well as the stability of liquid films; (b) *inserted films*, sandwiched between a rubber-like material and a solid, with applications to the process of adhesion onto a wetted solid surface and to the stability of the lachrymal film between a soft contact lens and the cornea; (c) *suspended viscous films* (in air or in a liquid with low viscosity); this configuration controls the stability of polymer and glass foams as well as that of emulsions.

In each of these cases, we will study the stability conditions of liquid films whose thickness ranges from a nanometer to a millimeter. We will also describe the dynamics of the dewetting process for a wide variety of liquids, ranging from water to ultra-viscous pastes, the viscosity of which is millions of times that of water. We will deal with three different dynamical regimes:

- "viscous regime," applicable to oils;
- "inertial regime," applicable to water;
- "viscoelastic regime," applicable to polymers and molten glass.

We will explain some rather unusual phenomena. As an example, dewetting at high speed can generate shock waves! Likewise, in liquid films millions of times more viscous than water, holes can open up at such high velocities (~m/s) that high-speed cameras are required to capture the phenomenon.

We will deal primarily with the dewetting of metastable films. We will also discuss the case of spinodal dewetting (of more restricted practical interest), which is limited to very thin films ($e < 10$ nm) and to ultra-pure conditions to avoid the nucleation of dry zones at specks of dust.

7.1   Critical Thickness for Dewetting

7.1.1   Film on a Solid Substrate

In a total wetting regime ($S > 0$), a film is always stable. By contrast, when $S < 0$, a film will dewet below a critical thickness e_c, as can be observed in a Teflon pan or on a plastic sheet. When you place a small drop of water on one of these materials, it will contract into a spherical cap. When you place a larger drop, it will form a puddle flattened by gravity. If you use

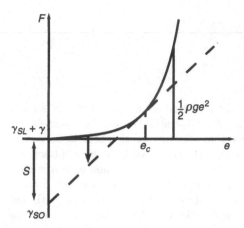

FIGURE 7.2. Free energy $F(e)$ as a function of the thickness e of a liquid film deposited on a non wettable substrate. Maxwell's construction describes the coexistence between the dry solid ($e = 0$) and a puddle ($e = e_c$), and shows that a film of thickness $e < e_c$ lowers its energy by breaking up into two phases (dry solid and puddle), as indicated by the arrow.

your finger to force that large drop to spread over a wider area, you will create an unstable film that dewets and breaks up. The conclusion of these simple observations is that the critical thickness e_c for dewetting is the puddle thickness discussed in chapter 2:

$$e_c = 2\kappa^{-1} \sin\left(\frac{\theta_E}{2}\right) \tag{7.1}$$

where κ^{-1} is the capillary length, of the order of a millimeter. To understand the threshold ec, we must consider the free energy $F(e)$ of a film as a function of its thickness, which is shown schematically in Figure 7.2.

Thick Films

For thick films, $F(e)$ is the sum of the interfacial energy and the gravitational energy:

$$F(e) = \gamma_{SL} + \gamma + \frac{1}{2}\rho g e^2 \tag{7.2}$$

where g is the acceleration of gravity and ρ is the density. In the absence of any film, when $e = 0$, we have $F(0) = \gamma_{SO}$.

$F(e \to 0)$ does not connect smoothly with $F(0)$. To describe this connection, it is necessary to include the long-range forces (see following section). Constructing Maxwell's double tangent (Figure 7.2) shows that, if $S < 0$ (i.e., if $\gamma_{SO} < \gamma_{SL} + \gamma$), there is an equilibrium between the dry solid and a film of thickness e_c (which corresponds to a flat puddle). The equilibrium is realized when one deposits a puddle of constant thickness on a dry solid. Any film with a thickness e less than e_c proves to be unstable; it will lower its energy by splitting up into two coexisting phases, namely, the dry solid and a puddle of thickness e_c. From equation (7.2), we can define a surface tension $\tilde{\gamma}_{SO}$ of the wet solid ($\tilde{\gamma}_{SO} = +\partial[F(e)\mathsf{A}]/\partial\mathsf{A}$, with $\mathsf{A}e$ = constant, where A is the wetted surface area). The quantity $\gamma_{\tilde{S}O}$ is then given by

$$\tilde{\gamma}_{SO} = \gamma_{SL} + \gamma - \frac{1}{2}\rho g e^2. \tag{7.3}$$

Equilibrium between the dry solid and the puddle can then be expressed as

$$\tilde{\gamma}_{SO} = \gamma_{SO}. \tag{7.4}$$

This equation is identical with the one obtained by expressing the equilibrium of forces acting on a portion of the puddle. Its solution is $e = e_c$. There is an analogy between a van der Waals pancake discussed in chapter 4 ($\tilde{\gamma}_{SO} = \gamma_{SL} + \gamma + e\Pi + P$, where Π is the disjoining pressure and P is the contribution of the van der Waals forces to the surface energy) and a gravity-driven pancake. Indeed, after defining $P_G = (1/2)\rho g e^2$ and $\Pi_G = -dP_G/de = -\rho g e$, we may write a similar expression $\tilde{\gamma}_{SO} = \gamma_{SL} + \gamma + e\Pi_G + P_G$. The quantity $\tilde{\gamma}_{SO}$ is the intersection of the tangent to the $F(e)$ curve with the y-axis. Maxwell's construction amounts to a graphical representation of the equilibrium condition $\tilde{\gamma}_{SO} = \gamma_{SO}$.

Thin Films

Equation (7.2) does not apply to microscopic films because $F(e \to 0)$ fails to merge with $F(0) = \gamma_{SO}$. To remedy this problem, we must include in $F(e)$ the contribution $P(e)$ of the long-range forces discussed in chapter 4. The complete expression for $F(e)$ then reads

$$F(e) = \gamma_{SL} + \gamma + P(e) + \frac{1}{2}\rho g e^2. \tag{7.5}$$

We shall restrict the analysis to the case of simple liquids, for which $P(e)$ is dominated by van der Waals forces:

$$P(e) = \frac{A}{12\pi e^2} \tag{7.6}$$

where $A = A_{SL} - A_{LL}$ is the Hamaker constant (negative in the present case) of the liquid on the solid.

Figure 7.3 depicts the energy $F(e)$ of a wetted solid for a wide range of liquid thicknesses, ranging from the microscopic to the macroscopic. We may deduce from the figure that, just as in the case of phase separation (segregation of binary mixtures, liquid-vapor phase transition), there are two mechanisms for the drop to recede.

- *For thick films*, where gravity dominates, the curvature of $F(e)$ is positive [$F''(e) > 0$] and the film is therefore metastable. Drying requires

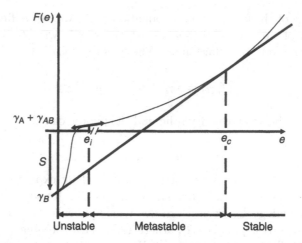

FIGURE 7.3. Two dewetting regimes: "spinodal" and "nucleation and growth." Depicted here is the energy $F(e)$ of a film whose thickness ranges from a nm to a mm, in a partial wetting regime, with the long-range forces included. Depending on the curvature of $F(e)$, one predicts two possible mechanisms for dewetting when $e < e_c$. If $F''(e) > 0$, the film is metastable and dewets by nucleation and growth of a dry zone; if $F''(e) < 0$, the film is unstable and breaks up spontaneously (spinodal decomposition). The thickness e_i at the inflection point is of the order of 10 nm.

the nucleation of a dry zone, which will expand if its radius exceeds a critical value $R_c \approx e$. This phenomenon has been studied both theoretically and experimentally.[2-4] It has been referred to as the *nucleation and growth* regime, first studied by C. Redon.[5]

- *For microscopic films*, the gravitational energy is negligible compared to the energy associated with long-range forces. The curvature of $F(e)$ becomes negative [$F''(e) < 0$] and the film is unstable. Capillary waves get amplified and the film spontaneously breaks up into a multitude of droplets arranged in a polygonal pattern. This regime has been dubbed *spinodal decomposition* by analogy with phase transitions. It was discovered by G. Reiter.[6] Such polygonal arrays of droplets can readily be created by spreading alcohol on a plastic sheet. As the alcohol evaporates, it leaves behind a microscopic film of water that breaks up spontaneously.

7.1.2   Film on a Liquid Substrate

Suppose we place a drop of oil A (for instance, octane) on a surface of water B kept in a vessel. The spreading coefficient $S = \gamma_B - (\gamma_A + \gamma_{AB})$ is negative, and the oil does not spread. Rather, it remains coalesced in a drop floating on the water's surface. As we keep adding more and more oil, lenses flattened by gravity (of thickness $e = e_c$) form and expand until the entire surface is covered by a single stable film of thickness $e > e_c$. At this point, suppose we siphon the oil out with a pipette and reduce the thickness of the oil film down to a value $e < e_c$. The film then finds itself in a metastable state. If we disturb it by blowing on it, instant dewetting takes place.

Here again, the thickness of floating lenses flattened by gravity is the critical thickness for dewetting, e_c. Concrete examples include a layer of crude oil spilled on the surface of the sea, or molten glass floating on liquid tin during the manufacture of floated glass. In the latter case, $e_c = 7$ mm, and it is a challenging problem to produce glass plates thinner than e_c (no surfactant is known in mineral chemistry to modify e_c). For octane on water, $e_c = 3.7$ mm.

The energy of a liquid substrate B covered with a floating film A is given by

$$F(e) = \gamma_{AB} + \gamma_A + P(e) + \frac{1}{2}\tilde{\rho}ge^2 \tag{7.7}$$

where $\tilde{\rho}$ is an effective density $\tilde{\rho} = \rho_A(1 - \rho_A/\rho_B)$, which includes Archimedes' force. Indeed, when a film of liquid A is floating, it displaces a thickness e' of liquid B such that $\rho_A e = \rho_B e'$. The gravitational energy is therefore given by $P_G(e) = (1/2)(\rho_A e^2 - \rho_B e'^2)$, or $(1/2)\tilde{\rho}e^2$. The critical thickness for dewetting is given by the equilibrium condition (bare liquid B / liquid covered by film A), which, by analogy with equation (7.3), reads

$$\gamma_B = \tilde{\gamma}_B = \gamma_{AB} + \gamma_A - \frac{1}{2}\tilde{\rho}ge_c^2. \tag{7.8}$$

This is equivalent to

$$-S = \frac{1}{2}\tilde{\rho}ge_c^2. \tag{7.9}$$

Since measuring angles on liquids is often difficult, e_c is determined directly by depositing increasingly large volumes Ω, measuring the radius R, and graphing the quantity $\Omega/\pi R^2$ as a function of $1/R$. Extrapolating the graph back to the origin gives the value of e_c [and the slope gives the edge energy $\Im$, as discussed in chapter 3, equation (3.38)].

To make a thick, metastable film, all we need to do is siphon out an initially stable film down to a thickness less than e_c. To make an even thinner film, one technique is to deposit a thin film of liquid A in solution in a solvent C that is wetting and volatile. Left behind after the evaporation of C is a thin, "unstable" pellicle of A.

7.1.3 Sandwiched Liquid Films

We now consider the case when a liquid film A deposited on a substrate B (solid or liquid) is exposed not to air but to a deformable medium R, which can be either another liquid or a soft solid (an elastomer). The spreading parameter S is

$$S = \gamma_{BR} - (\gamma_{AB} + \gamma_{AR}). \tag{7.10}$$

The energy per unit surface of a microscopic film is given by

$$F(e) = \gamma_{AB} + \gamma_{AR} + P(e). \tag{7.11}$$

For a thick film, one must add to equation (7.11) a gravity term $P_G = (1/2)(\rho_A - \rho_R)ge^2$. $P(e)$ is a quadratic function of the polarizabilities α_i, which must vanish when $A \equiv B$ or $A \equiv R$. Therefore, it has the functional form

$$P(e) = k(e)(\alpha_A - \alpha_B)(\alpha_A - \alpha_R). \tag{7.12}$$

If $B = R =$ vacuum, we have $P(e) = -A_{AA}/12\pi e^2$, which defines $k(e)$. From this we deduce

$$P(e) = \frac{A}{12\pi e^2} \tag{7.13}$$

where the coefficient A is defined by $A \equiv -(A_{AA} - A_{AR} - A_{BA} + A_{BR})$. When $R \equiv 0$ (air) and B is a solid, we recover $A = A_{SL} - A_{LL}$, as discussed in section 7.1.1. Substituting a medium "R" for air allows us to adjust the Hamaker constant A and, hence, the stability of thin films.[6-8] If $A > 0$, the resulting thin film will be stable. On the contrary, if $A < 0$, it will be unstable (this is always the case with suspended films, for which $B \equiv R \equiv 0$). We will discuss in this chapter how suspended films recede. The case of sandwiched films when medium R is an elastomer will be treated in chapter 9.

7.2   Viscous Dewetting

In this section, we are interested in the dewetting of viscous liquids, for which inertial effects are negligible. The dynamics of the phenomenon will then be controlled by the conversion of surface energy and gravity (perhaps even of van de Waals energy in the case of very thin films) into viscous dissipation. The inertia of the ridge is negligible. The range of viscosities we will deal with goes from roughly one mPa-s for water to as high as 10^6 mPa-s for ultra-viscous PDMS.

We begin by describing a table-top experiment (Figure 7.4) that is straightforward to set up with a viewgraph projector, a transparency sheet, sulfochromic acid, colored water, glycerol (to increase the viscosity of water), and a video camera. Sulfochromic acid attacks plastic and renders it wettable. The first step is to dip the rim of a drinking glass into the acid, to place it briefly on the transparency sheet, and then to rinse the sheet thoroughly with tap water. This procedure produces a wetting ring that will pin a film of water containing both glycerol to slow down the dewetting process and a dye to improve the contrast.

The experiment itself consists of depositing a small puddle of water at the center of the ring and to spread it with a circular rod (a pencil works nicely). This produces a circular film that fills the inside of the wetting ring. The film formed in this way is metastable. With the help of a pipette

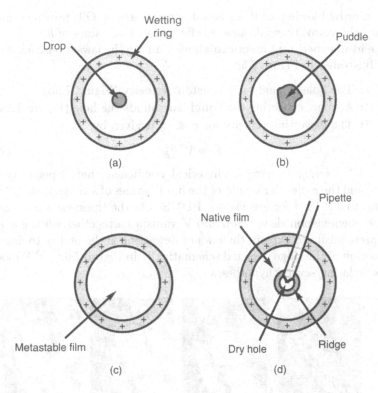

FIGURE 7.4. Experimental demonstration of dewetting: A drop (a) or a puddle (b) is spread as a film inside a wetting ring (c). The liquid is removed by suction to nucleate a dry zone that subsequently expands (d).

or a piece of blotting paper, one can create a hole in the center of the ring. Beyond a critical radius, the hole enlarges and its growth can be recorded with the camera. Careful examination reveals that the hole is encircled by a ridge collecting the liquid. If its size is monitored at regular intervals, its radius $R(t)$ turns out to depend linearly on time. In other words, the dry zone expands at a velocity $V = dR/dt$ that is constant.

7.2.1 Ideal Solid Substrates

C. Redon conducted the first quantitative dewetting experiments.[5] They involved alkanes deposited on nearly ideal surfaces prepared by J. B. Broszka according to J. Sagiv's method.[9, 10] The substrate preparation technique is based on depositing a compact molecular coating of fluorinated or hydrogenated silanes (see chapter 1) on top of a silicon wafer that is smooth on an atomic scale. The periphery of the wafer is made wettable by locally destroying the molecular coating with a laser beam. PDMS films can readily be spin-coated on a wafer. A hole in the film can be created either by suction or by blowing air if the liquid is very viscous. Oils provide a means to vary the viscosity, while alkanes offer a range of values of θ_E.

The ideal experiment described above leads to the laws of viscous dewetting illustrated in Figure 7.5:

Law 1: The hole expands at a constant velocity (Figure 7.5b).
Law 2: A ridge, collecting the liquid, surrounds the hole (Figure 7.5a).
Law 3: The dewetting velocity for $e \ll e_c$ is given by

$$V = V^*\theta_E^3 \tag{7.14}$$

where $V^* = k\gamma/\eta$, k being a numerical coefficient that depends on the nature and the molecular weight of the fluid. Values of k range from 10^{-2} for alkanes to 3×10^{-3} for less viscous PDMSs. As the thickness e approaches e_c, the phenomenon slows down and V vanishes altogether when $e = e_c$.

A prerequisite to explain the laws of dewetting is the ability to describe the motion of the ridge depicted schematically in Figure 7.6.[11-13] We begin by formulating several hypotheses:

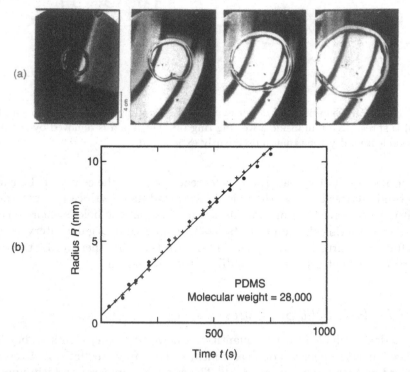

FIGURE 7.5. (a) Photographs of the growth of a dry zone in a PDMS film of molecular mass 28,000 deposited on fluorinated silicon. The total height of the image is 4 cm and the film thickness is 30 μm. (b) Time dependence of the radius $R(t)$ of the expanding dry hole. (From C. Redon, F. Brochard, and F. Rondelez, In *Physical Review Letters, 66* p. 715 (1991), © Americal Physical Society. Reproduced by permission.)

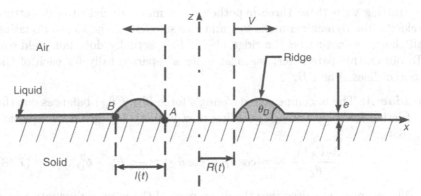

FIGURE 7.6. Schematic diagram of a film's profile during dewetting. A circular ridge is bounded by the contact line A, which recedes on the dry solid, and by point B, which advances on the wet solid.

1. The cross section of the ridge is circular. The pressure comes to equilibrium rapidly in the central portion. Hence, in accordance with Laplace's law, the curvature is constant. Accordingly, the profile of the rim is a circular arc that intersects the horizontal with the same dynamical angle θ_D at points A and B.

2. The ridge collects the liquid that initially filled the hole. Because of conservation of volume, the size l of the ridge is related to the radius R of the hole:

$$\pi R^2 e = \pi R l^2 \theta_D. \tag{7.15}$$

3. The viscous dissipation is dominated by the fluid's flow within the ridge (hydrodynamic model). The motion of the ridge results from an equilibrium between two forces:

   i. The first is the driving force F_M, given by

   $$F_M = \gamma_{SL} + \gamma - \gamma_{SO} - \frac{1}{2}\rho g e^2 = \frac{1}{2}\rho g(e_c^2 - e^2). \tag{7.16}$$

   It follows that the ridge is pulled by a constant force. In the limit of small contact angles ($\theta_E \ll 1$) and small thicknesses ($e \ll e_c$), the gravity term $(1/2)\rho g e^2$ can be neglected and F_M reduces to

   $$F_M = \frac{1}{2}\gamma\theta_E^2. \tag{7.17}$$

   ii. The second is the friction force F_V exerted by the solid on the liquid in motion. The viscous dissipation is dominated by friction in the liquid wedges that form the boundaries of the ridge, and this force does not depend on the size of the ridge. F_V is proportional to the velocity, and imposing $F_V = F_M$ thus leads to a dewetting velocity that is constant in time (law 1).

Starting with these three hypotheses, we may calculate the dewetting velocity, the dynamical angle θ_D, and the size $l(t)$ of the ridge. Equating all the forces acting on the ridge ($F_V = F_M$) actually does not yield θ_D. To obtain this parameter, we must write a separate tally for each of the contact lines A and B.

- **Line A:** The uncompensated Young's force ($\theta_D \neq \theta_E$) balances out the friction force on a wedge receding at a velocity V_A [given by equation (6.4)]:

$$\frac{3\eta V_A \ln_A}{\theta_D} = \gamma(\cos\theta_D - \cos\theta_E) \approx \frac{1}{2}\gamma(\theta_E^2 - \theta_D^2). \qquad (7.18)$$

  The quantity $\ln_A$ describes the divergence of the viscous dissipation in a wedge [see equation (6.5)]. It is written as $\ln_A = \ln(l/a)$, where a is the molecular size. In practice, $\ln_A$ is in the range of 10 to 20.[13]

- **Line B:** The wedge advances by gliding on the liquid film covering the solid. As such, the equilibrium angle should be $\theta_E = 0$. Equilibrium of the forces acting on the line B, which advances at a velocity V_B, can thus be written as

$$\frac{3\eta V_B \ln_B}{\theta_D} = \gamma(1 - \cos\theta_D) - \frac{1}{2}\rho g e^2 \approx \frac{1}{2}\gamma\theta_D^2 \qquad (7.19)$$

  which holds in the limit of $e \ll e_c$. Here, we have $\ln_B = \ln(l/e)$ since the wedge glides on the wet solid ($\ln_B \approx 1$).

Since dl/dt is small compared to the velocity of the ridge, we can use the approximation $V_A = V_B$. This condition fixes the value θ_D:

$$\theta_D = \frac{\theta_E}{(1 + \ln_A / \ln_B)^{1/2}} \approx \frac{\theta_E}{4}. \qquad (7.20)$$

This value is distinctly smaller than the estimate arrived at by early workers, who proposed the expression $\theta_D = \theta_E/\sqrt{2}$ based on the assumption $\ln_A = \ln_B$.[11,12] C. Andrieu used the grid technique (discussed in chapter 2) to measure θ_D and was able to confirm the prediction of equation (7.20), namely, that θ_D is indeed proportional to θ_E, but is less than $\theta_E/\sqrt{2}$.[14]

Substituting the value of θ_D back into equation (7.19), and in the approximation $e \ll e_c$ (law 3), we get

$$V = V^*\theta_E^3 \qquad (7.21)$$

where $V^* = \gamma/6\eta \ln$, the quantity "ln" being a logarithmic coefficient related to $\ln_A$ and $\ln_B$. Since geometrical factors enter only via their logarithms, we conclude that V is virtually independent of e in the limit $e \ll e_c$.

When e is no longer negligible in comparison to e_c, we expect two distinct regimes:

1. *For short times*, the ridge retains its circular shape. The force of friction remains unaltered, but the driving force decreases as the film becomes thicker [equation (7.16)]. The sum of equations (7.18) and (7.19), in which we keep the gravity term, gives the equilibrium of the overall forces:

$$\frac{6\eta V \ln}{\theta_D} = \frac{1}{2}\gamma\theta_E^2 - \frac{1}{2}\rho g e^2 = \frac{1}{2}\rho g(e_c^2 - e^2). \tag{7.22}$$

The hole starts expanding at a constant velocity V, but that velocity slows down as the film gets thicker because the driving force F_M ultimately decreases and actually vanishes altogether at $e = e_c$.

2. *For long times*, the ridge eventually flattens, and we enter the gravity regime.[15] The viscous dissipation within the liquid increases and can overcome the dissipation in the wedges. The dewetting process slows down as time progresses. At point A, θ_D approaches θ_E and the ridge becomes a ribbon of length l and thickness $e = e_c$.* The viscous dissipation can be written as

$$3\eta \left(\frac{V}{e_c}\right)^2 le_c = F_V V. \tag{7.23}$$

The friction force deduced from equation (7.23) increases linearly with the size l of the ridge. Equating F_M and F_V gives

$$3\eta \left(\frac{l}{e_c}\right) V = \frac{1}{2}\rho g(e_c^2 - e^2). \tag{7.24}$$

The conservation of volume written for a ribbon of thickness e_c reads

$$2\pi R l(e_c - e) = \pi R^2 e. \tag{7.25}$$

Equations (7.24) and (7.25) taken together lead to the time-evolution law for $R(t)$:

$$R(t)^2 = Dt \tag{7.26}$$

where $D \approx V^* e_c \kappa^2 (e_c - e)^2$. The thermodynamic drop-off in velocity ($V \to 0$ when $e \to e_c$) described in equation (7.22) can indeed be observed experimentally, but the "diffusive" regime embodied in equation (7.26), which requires large surfaces (of the order of a meter), has yet to be studied.

---

*Actually, the flow is associated with a pressure gradient created by a slight slope $\rho g dz/dx$.

Conclusion. The dewetting process of metastable films takes place by nucleation and growth of a dry hole. The kinetic properties of dewetting results from a balance between the energy gained per unit time $(2\pi RSV)$ and the energy dissipated in the moving ridge $(\approx 2\pi R\eta V/\theta_E)$. For thin films $(e \ll e_c)$, the dewetting velocity obeys the law $V = V^*\theta_E^3$, which validates the hydrodynamic model.

Having described the ideal case, we proceed next to examine situations more in keeping with practical applications.

7.2.2   Imperfect Solid Substrates

7.2.2.1   Surfaces With Hysteresis

In practice, solid surfaces typically exhibit hysteresis effects. Drops remain pinned instead of rolling as they would on silanized silicon wafers. There is no longer a single contact angle θ_E, but a whole range of values between a receding angle θ_R and an advancing angle θ_A. As we have seen in chapter 3, this hysteresis phenomenon has two possible origins—chemical defects and surface roughness.

C. Andrieu et al. have studied the dewetting of "dirty" surfaces, in particular, films of mylar (transparencies) and of Teflon, both of which exhibit a strong hysteresis.[16] L. Bacri has studied the dewetting of porous, hydrophobic substrates with very rough surfaces.[17] The first question we will address is what becomes of the critical dewetting thickness e_c on such surfaces.

Puddles on "Dirty" or Rough Surfaces

On any hysteresis-prone surface, there are two extreme contact angles—θ_A and θ_R. For a puddle, depicted in Figure 7.7a, we will therefore measure an advancing thickness e_A (the puddle connects with the solid at an angle θ_A) and a receding thickness e_R (the puddle connects with the solid at an angle θ_R). These thicknesses are given by

$$e_A = 2\kappa^{-1}\sin\frac{\theta_A}{2}; \qquad e_R = 2\kappa^{-1}\sin\frac{\theta_R}{2}. \tag{7.27}$$

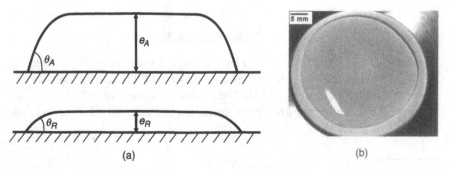

(a)   (b)

FIGURE 7.7. (a) Advancing and receding thicknesses of a puddle placed on a dirty surface. (b) Puddle on a porous, hydrophobic membrane (courtesy L. Bacri).

Shown in Figure 7.7b is a puddle of water placed on a porous membrane exhibiting a strong hysteresis. Measuring e_A and e_R is done simply by recording the radius of the puddle at a given volume. For the example at hand, and with the help of equations (7.27), we deduce the values $\theta_A = 90°$ and $\theta_R = 21°$.

Suppose now that we deposit a thick film, the thickness of which we gradually reduce by suction with a pipette (or with a pump if we are dealing with a porous material). A hole created in the film will close up when $e > e_A$. When $e_R < e < e_A$, a hole will remain stable since the angle θ associated with e via equation (7.1) is between θ_R and θ_A and the contact line can neither advance nor recede. When $e < e_R$, a hole will open up spontaneously since $\theta < \theta_R$. It therefore follows that the critical thickness for dewetting on a dirty or rough surface is e_R.

Dynamics of Dewetting with Hysteresis

The dynamical laws of dewetting are roughly the same as in the "ideal" case as long as we replace θ_E by θ_R.[16] The velocity with which a hole opens up is constant and given by

$$V = V^*\theta_R^3$$

which holds for thin films. V approaches zero when $e \to e_R$ in accordance with

$$V = V^*\kappa^3(e_R^2 - e^2)e_R. \tag{7.28}$$

Note:
Bacri has studied the case of a film with a pre-made hole $(e < e_A)$ deposited on a soaked, porous substrate being drained at a constant flow rate J (Figure 7.8).[17] The hole remains stable as long as $e > e_R$. When $e = e_R$, dewetting begins. If we adopt this particular moment as the origin of time (which implies $e = e_R - Jt$), equation (7.28) gives the time dependence of the dewetting velocity V:

$$V = 2V^*\kappa^3 e_R^2 J \cdot t$$

which holds as long as t is less than the suction time $t_S = e_R/J$.

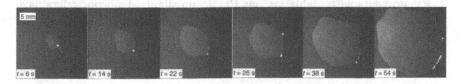

FIGURE 7.8. Dewetting of a film of water deposited on a rough and porous surface (courtesy L. Bacri).

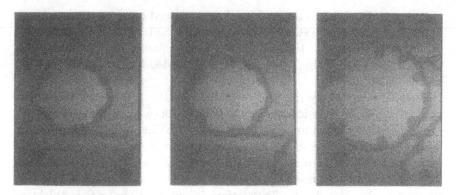

FIGURE 7.9. Opening of a hole in a 400-μm thick film. (From C. Redon, J. B. Brzoska, and F. Brochard. In *Macromolecules, 27*, p. 468 (1994), © 2001 American Chemical Society. Reproduced by permission.)

7.2.2.2 "Slippery" Substrates

Redon has studied the dewetting of ultra-viscous PDMS on ideal surfaces (Figure 7.9). She showed that on such smooth and passive surfaces, a highly viscous fluid slides.[18]

The sliding is characterized by an extrapolation length b.[19] For polymers, $b \approx aN^3/N_e^2$, where N is the degree of polymerization, N_e is the threshold for entanglement, and a is the size of the monomer. Taking $N = 10^3$, $N_e = 10^2$, and $a = 3$ Å, we find $b = 3$ μm. If the film thickness $e \gg b$, the film flows like an ordinary liquid. By contrast, if $e < b$, the polymer moves in unison like a solid and all the dissipation occurs at solid/liquid interface. This type of fluid transport is called a *plug flow*.*

When the ridge of size l slides on the solid with a velocity V, the friction force (again per unit length) is

$$F_V = klV$$

where $k \approx \eta_0/a$ is the coefficient of friction (η_0 is the viscosity of a solution of monomers of size a). Balancing this force with the driving force $|S|$, and taking into account the conservation of volume $[eR \approx l^2\theta_E$, equation (7.15)], we find

$$R \propto t^{2/3} \tag{7.29}$$

Experiments with highly viscous PDMS on silanized silicon wafers have shown that, for thick films ($e > 10$ μm), the dewetting is of the classic type and obeys the law $R \propto t$. On the contrary, for microscopic films ($e < 1$ μm), C. Redon found a dependence of the type $R \propto t^{0.66\pm0.01}$. Such a law is the signature of a sliding polymer. It has subsequently been observed during

---

*In this type of flow, there is no lateral velocity gradient. Instead, the fluid moves as a unit, much like a cork being extracted from a bottle.

the dewetting of polymer films deposited on lubricated solids, that is to say, on substrates covered with a thin film of wetting liquid with low viscosity, or a molecular coating.[20]

In summary, when the ridge slides, the friction force increases with its size l because all the viscous dissipation is concentrated at the S/L interface. The ridge slows down as time progresses, and the law $R \propto t^{2/3}$ describing the rate of growth of the hole is characteristic of a sliding mode of motion.

7.2.3 Liquid Substrates

Liquid substrates, smooth and homogenous, are the epitome of perfection. However, they deform and flow. The challenge is to take into account the flows induced by the motion of the ridge.

A. Buguin and P. Martin were the first to conduct liquid-on-liquid dewetting experiments.[21] They deposited PDMS (polydimethylsiloxanes) on the fluorinated oil PFAS (polyfluoromethylalkylsiloxane), which is an oligomer, the viscosity of which depends on the length of the chain. They measured a critical thickness $e_c = 1.3$ mm on puddles. This result implies $S = -1.8$ mN/m and $\gamma_{AB} = 3.5$ mN/m.

Films of thickness e can be created by depositing on the PFAS substrate a solution of PDMS in a wetting and highly volatile solvent (isopentane). The delicate part of the experiment is the evaporation of the solvent, which is susceptible to Marangoni-Bénard instabilities likely to induce dewetting (see Figure 7.10a). Once this hurdle is overcome, one ends up with a pure, metastable film floating on the surface.

To create a hole without perturbing the fluid substrate, Buguin hit on the idea of injecting a bubble with a pipette. After rising gently to the

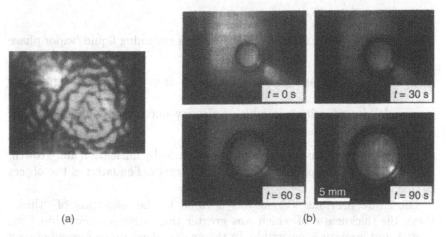

(a)   (b)

FIGURE 7.10. (a) Picture of the surface during the evaporation of the solvent, which creates Marangoni flows; (b) dewetting of a PDMS film deposited by evaporation on a PFAS pool. (From P. Martin, A. Buguin, and F. Brochard, in *Europhysics Letters*, *28*, p. 421 (1994), © EDP Sciences. Reproduced by permission.)

surface, the bubble bursts, initiating a hole that subsequently opens up (Figure 7.10b). The perfectly circular hole is surrounded by a ridge collecting the liquid film. It expands at a constant rate that proves independent of the film's viscosity (at least as long as it is not too high, i.e., when $\eta_A < \eta_B R/e$):

$$V \approx \frac{|\tilde{S}|}{\eta_B} \qquad (7.30)$$

where $|\tilde{S}| = (1/2)\rho g(e_c^2 - e^2)$ is the driving force acting on the ridge and η_B is the viscosity of the substrate.

As the ridge (of size l) advances by floating on the liquid at a velocity V, it induces flows, and all the viscous dissipation takes place within the substrate. Since the penetration length of the flow in the B liquid substrate is l, the viscous dissipation can be written as

$$F_V V \approx \eta_B \left(\frac{V}{l}\right)^2 l^2 \qquad (7.31)$$

which reduces to

$$F_V \approx \eta_B V. \qquad (7.32)$$

We have just rediscovered Stokes' formula for a cylinder, except for a logarithmic coefficient discussed in the literature.[22] Equilibrium of the forces $|\tilde{S}| = F_V$ acting on the ridge does indeed lead straight to the law expressed in equation (7.30).

In summary, dewetting of a macroscopic film supported by a liquid proceeds by nucleation and growth of a hole that expands at a constant velocity V. This velocity is independent of the viscosity of the film but does depend on that of the substrate through the relation $V \approx |\tilde{S}|/\eta_B$.

7.2.4 Spinodal Dewetting

There exist two phase separation mechanisms governing liquid/vapor phase transitions:

- Nucleation at impurities and growth of drops if the vapor phase is metastable;
- spinodal decomposition in which a density fluctuation gets amplified if the vapor phase is unstable.

The first mechanism corresponds to dewetting by nucleation and growth, the second to the amplification of capillary waves. The latter is the object of this paragraph.

The entire previous section was devoted to the dewetting of "thick" films, the thickness e of which was greater than a micron. For thin films ($e \ll 1$ μm), gravity is negligible. In this regime, long-range forces between liquid and substrate dominate. They can be written as

$$F(e) = \gamma_{SL} + \gamma + P(e). \qquad (7.33)$$

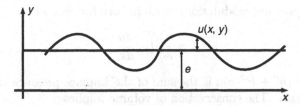

FIGURE 7.11. Thickness fluctuations of a very thin wetting film.

We consider the case of van der Waals forces, for which we have $P(e) = A/12\pi e^2$.

If a film of average thickness e exhibits a modulation $u(x, y)$ (see Figure 7.11), the corresponding increase in energy per unit volume reads

$$F - F_0 = \frac{1}{2}\gamma|\nabla u|^2 + \frac{1}{2}P''(e)u^2 \qquad (7.34)$$

where $P''(e) = A/2\pi e^4$.

The first term is associated with the surface tension of the liquid. It is simply proportional to the increase in area of the free surface calculated in the approximation of a small modulation ($|\nabla u| \ll 1$). The second term is the van der Waals contribution. Comparing these two terms defines a characteristic length ξ_e, called the *healing length*, which was discussed in chapter 4:

$$\xi_e = e^2 \left(\frac{2\pi\gamma}{|A|}\right)^{1/2} \approx \frac{e^2}{a} \qquad (7.35)$$

where a is a molecular dimension. As an example, for $e = 100$ Å, the healing length comes out to be of the order of 1 μm.

If A is positive, the film is stable. When deposited on a wavy substrate characterized by a wavevector q, the film replicates the surface's modulation if $q\xi_e < 1$, whereas it remains planar if $q\xi_e > 1$ (see chapter 4). That is the reason why ξ_e has been dubbed the "healing length," as the film "heals" the roughness of the original surface.[23]

We now focus our attention on the case $A < 0$, which implies an unstable film. Equation (7.34) tells us that, in this instance, the energy of the film decreases when the thickness is modulated with a spatial periodicity q^{-1} greater than ξ_e. If $q\xi_e < 1$, the thickness modulation gets amplified in time, which leads to spinodal dewetting.

We examine next the dynamical properties of the process for short times, in the linear regime, and for small amplitudes ($u \ll e$). We assume a modulation of wavevector q in the x-direction. The thickness modulation $u(x, t)$ is

$$u(x, t) = u_0 \exp(iqx) \exp\left(\frac{t}{\tau}\right). \qquad (7.36)$$

It creates a pressure modulation, which in turn induces a flow J in the film:

$$J = \frac{e^3}{3\eta}\left(-\frac{dp}{dx}\right) \tag{7.37}$$

where $p = -\gamma u'' + P''(e)u$ is the sum of the Laplace pressure and the disjoining pressure. The conservation of volume implies

$$\frac{\partial u}{\partial t} + \nabla J = 0. \tag{7.38}$$

Upon replacing u by its expression given in equation (7.36), the last two equations taken together give the time constant τ of the fluctuation:

$$\frac{1}{\tau} = -\frac{q^2 e^3}{3\eta}(\gamma q^2 + P''(e)) = -V^* e^3 q^2 \left(q^2 - \frac{1}{\xi_e^2}\right) \tag{7.39}$$

where $V^* = \gamma/3\eta$. If $q\xi_e > 1$, τ is negative, and the fluctuation dies out. If $q\xi_e < 1$, τ is positive and the fluctuation grows with time. The wavevector q_M of the fastest mode is obtained by maximizing equation (7.39). The result is:

$$q_M = \frac{1}{\sqrt{2}}\frac{1}{\xi_e}. \tag{7.40}$$

The corresponding time constant τ_M is:

$$\tau_M = \frac{4\xi_e^4}{e^3 V^*} = \tau_0\left(\frac{e}{a}\right)^5$$

where τ_0 is a microscopic time given by $\tau_0 \approx a/V^*$.

Note:
Equation (7.39) can also be derived from energy considerations. For a mode of wavevector q and amplitude u_q, equation (7.34) gives

$$F_q = \frac{1}{2}\gamma\left(q^2 - \frac{1}{\xi_e^2}\right)u_q^2. \tag{7.41}$$

We can estimate the dispersion relation by writing that the energy gained per unit time is converted into a viscous dissipation $T\dot{S}$:

$$\dot{F}_q + T\dot{S} = 0. \tag{7.42}$$

The quantity $T\dot{S}$ can be calculated exactly for a Poiseuille flow. In a simple dimensional analysis, we have (per unit surface area)

$$T\dot{S} \propto \eta\frac{v_x^2}{e^2}e. \tag{7.43}$$

The conservation of volume provides a relation between v_x (velocity in the x-direction) and $\dot{u}$:

$$ev_x \propto \dot{u}q^{-1}. \tag{7.44}$$

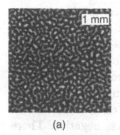
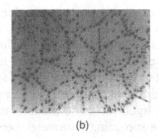

(a)   (b)   (c)

FIGURE 7.12. Images of the spinodal dewetting of a nanoscopic film of PDMS deposited on a silicon wafer (a) for short times, showing the amplification of the thickness fluctuations (courtesy O. Rossier); (b) for long times (courtesy G. Reiter); (c) also for long times, but on a liquid substrate (courtesy A. Buguin).

Equation (7.42) can then be recast in the form

$$\gamma \left(q^2 - \frac{1}{\xi_e^2} \right) u\dot{u} \propto \eta \frac{\dot{u}^2}{e^3 q^2} \tag{7.45}$$

which leads straight to equation (7.39).

The first observations of spinodal dewetting are due to G. Reiter.[6] He used optical microscopy to study the dewetting of a polystyrene film ($e < 20$ nm) deposited on a silicon wafer. He observed that the film became unstable (Figure 7.12a) and broke up into a multitude of droplets arranged in a polygonal pattern (Figure 7.12b). Fourier analysis of the image revealed that the characteristic length of the polygons was $\xi_e \approx e^2/a$ and that the size of the drops obeyed the law $r \propto e^{3/2}$.

This result can be interpreted as follows. The long-wavelength ($\lambda > \xi_e$) thickness fluctuations get amplified and the film becomes wrinkled (Figure 7.12a). The thinnest regions nucleate a hole, which expands by forming a ridge. When two ridges meet, the radius of the hole is typically of the order of ξ_e and the lateral dimension l of a ridge is such that $l^2\theta_e \approx \xi_e e$. Because of the Rayleigh instability, a ridge is unstable and spontaneously fragments into droplets of size $l \propto \sqrt{\xi_e e} \propto e^{3/2}$, as shown in Figure 7.12b.

Since then, spinodal dewetting at short times has been studied in some detail. It has been possible to monitor by AFM the thickness modulation and its amplification as a function of time.[24-26] The literature contains discussions of the behavior of other materials.[27,*] Nevertheless, results are typically poorly reproducible and the appearance of holes at short times (observed in ultra-thin polystyrene films less than 10 nm thick) is often unexplained. These anomalies are usually blamed on flawed preparations

---

*In the case of liquid crystal films, for which the orientation of molecules is fixed at the walls, the distortion energy of the molecular axis can dominate over the van der Waals energy, and spinodal dewetting can take place in thicker films, which makes it easier to observe (see reference[27]).

of films by spin coating or by solvent evaporation.[6] Since the polymer used is generally vitreous at ambient temperature, the rapid evaporation of the solvent is bound to leave behind polymer chains in a frozen state. When subjected to mechanical stresses, such a film might evolve in a more complex way than predicted by the normal laws of spinodal dewetting.

P. Martin and A. Buguin have observed the dewetting of an ultra-thin PDMS film deposited on a liquid substrate, a fluorinated oil (PFAS). Dewetting begins as soon as the spreading parameter becomes negative. There always remains a small amount of solvent when S becomes negative, which creates Marangoni effects during dewetting. The film breaks up spontaneously into droplets of uniform size organized in a lattice (Figure 7.12c). The dynamics is much quicker on a liquid substrate, but studying it has so far been thwarted by the parasitic Marangoni effect. Overlooking this complication, the viscous dissipation in the liquid substrate B dominates.[28] Equation (7.43) then becomes

$$T\dot{S} \propto \eta_B \frac{v_x^2}{l_P^2} l_P \qquad (7.43b)$$

where l_P is the penetration length of the flow, given by $l_P^{-1} = \sqrt{q^2 + i\omega\frac{\rho}{\eta}} \propto q$.

In a purely viscous regime, the variation in energy remains unchanged, and the dispersion equation is

$$\frac{1}{\tau} = \frac{e^2\gamma}{2\eta_B}\left(q^2 - \frac{1}{\xi_e^2}\right)q. \qquad (7.46)$$

The time constant of the instability corresponding to the fastest mode ($q_M \approx 1/\xi_e$) is now given by $\frac{1}{\tau_M} = \frac{\gamma}{\eta_B}\frac{e^2}{\xi_e^3}$, or $\tau_M = \tau_0 \left(\frac{e}{a}\right)^4$.

7.3   Inertial Dewetting

We now consider the case of dewetting at high velocity, when the dynamics is controlled by the rate at which surface energy is being converted to kinetic energy of the liquid.

One might think of superfluid He-4 as the ideal candidate material. By virtue of its extremely small polarizability, liquid helium spreads on virtually any solid. One exception identified recently is cesium, which is subject to partial wetting in certain temperature ranges. Cesium thus qualifies as a good substrate material to study dewetting phenomena, and under the conditions just described, dewetting ought to be inertial. But the process is probably complicated by the nucleation of vortex lines.

As it happens, the simplest material combination available for studying the inertial regime is water on highly hydrophobic surfaces, in which case dewetting can proceed at ultra-fast rates, with velocities approaching 10

FIGURE 7.13. Inertial dewetting on hydrophobic glass. Ripples are emitted around the hole. (From F. Brochard, E. Raphël, and L. Vovelle, in *Comptes Rendus de l'Académie des Sciences, 321*, p. 367 (1995); Reproduced by permission.)

cm/s. Quickly drying wet surfaces has important industrial applications. As a result, a number of detailed studies have been conducted recently on this topic, which had until then been largely ignored.

C. Andrieu was the first to examine the dewetting of water on plastic.[14] One conclusion of this study was that, upon decreasing the thickness e, the velocity diverges as $1/\sqrt{e}$ instead of approaching a limiting value independent of e as would happen in a viscous regime.

Using a high-speed camera, L. Vovelle was able to observe the appearance of ripples. As a hole opens up, it emits capillary waves (Figure 7.13).[29]

A. Buguin conducted the first quantitative study of the dewetting of water on highly hydrophobic silanized glass.[30] With the help of a very simple optical setup, he was able to follow the time evolution of the profile of a water film. He identified several distinct regimes, which we proceed to describe in the next section.

Two dimensionless numbers play a key role in inertial dewetting phenomena:

1. The Reynolds number, which distinguishes between inertial and viscous regimes; and
2. the Froude number, which is associated with the appearance of ripples.

7.3.1 The Reynolds Number

The Reynolds number compares inertia and viscosity. It is defined by

$$R_e \equiv \frac{Vl}{\nu} \qquad (7.47)$$

where l is the size of the ridge and ν is the kinematic viscosity of the liquid ($\nu = \eta/\rho = 10^{-6}$ m^2/s for water). Using the values $l = 1$ mm and $V = 0.1$ m/s, we find $R_e = 100 \gg 1$.

Whenever the Reynolds number is large, the equation of motion of a ridge is fundamentally altered. Consider the case of a one-dimensional ridge. The equation of motion for a liquid mass $M(t)$ contained within the ridge is

$$\frac{d[M(t)V]}{dt} = F_M - F_V \tag{7.48}$$

where $F_M = \gamma_{SO} - \tilde{\gamma}_{SO} = -\tilde{S} = -S - (1/2)\rho g e^2 = (1/2)\rho g(e_c^2 - e^2)$, and F_V is the force of friction.

When $R_e \ll 1$, dewetting is viscous and we then have $F_M = F_V$. This limit corresponds to the analysis in section 7.2 where we have neglected the inertial term $d(MV)/dt$. When $R_e \gg 1$, dewetting is inertial and F_V is negligible. The second term in equation (7.48) no longer depends on V. The ridge of mass $M(t) = \rho e R$ advances at a constant velocity V that is the solution of equation (7.48). That velocity is

$$V = \sqrt{\frac{|\tilde{S}|}{e\rho}} \tag{7.49}$$

This law is similar to the one derived by F. Culick, which describes the bursting of soap films, after substituting 2γ for $|\tilde{S}|$ (see section 8.4.4).[31]

The same law can also be derived from energy arguments. The surface energy $|\tilde{S}|R$ gained per unit length of the ridge is converted into kinetic energy. For a ridge moving at a constant velocity while picking up weight, the kinetic energy is

$$E_K = \int F \, dx = \int \frac{d}{dt}(MV^2) \, dt = MV^2. \tag{7.50}$$

Taking into account the conservation of overall mass, expressed as $M = e\rho R$, $E_K = |\tilde{S}|R$ leads directly to equation (7.49) for the velocity. However, equation (7.50) is quite surprising. The kinetic energy of the ridge at a given time t is $\frac{1}{2}MV^2$. The conclusion is that exactly half the energy is converted into heat! The exact nature of the dissipation mechanisms remains mysterious, as is the bursting of a soap film. The great Lord Rayleigh himself erred by a factor of 2 when he wrote down the conservation of energy and omitted the disappearance of half the surface energy in the form of heat.[32] The situation can be compared to the soft impact of a ball that remains stuck on a target instead of bouncing back. Likewise, the ridge arrives with a velocity V as it enters a liquid film at rest.

Figure 7.14 shows velocities recorded with a video camera as a hole opens up in water during dewetting on hydrophobic glass. The hole grows at a velocity V and the front advances at a velocity V_F on the stationary liquid film. For small thicknesses, the ridge remains thin and $V = V_F$, but as the thickness increases, the ridge enlarges and becomes flattened by gravity, and $V < V_F$.

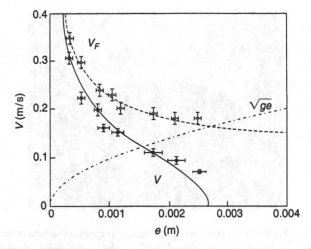

FIGURE 7.14. Dewetting velocity V (solid line) and advancing velocity V_F of the front (dashed line) as functions of the thickness of a film of water. Also shown (dash-dot) is the velocity of capillary waves in shallow water [equal to $(ge)^{1/2}$] (courtesy A. Buguin).

Culick's law [equation (7.49)] is indeed verified ($V \propto 1/\sqrt{e}$), but the dynamical value $|\tilde{S}_d|$ deduced from the experiment falls short of the static value by approximately 50%. Several mechanisms could be invoked to account for slowing the dewetting process (e.g., viscous losses in the boundary layer, "wave resistance" induced by the propagation of capillary waves, radial flow in the ridge, and non-uniform velocity assumed in a one-dimensional analysis, etc.).

7.3.2   The Froude Number (Condition for Shock Waves)

This number is defined by

$$F_r \equiv \frac{V_F}{\sqrt{ge}} \tag{7.51}$$

where V_F is the velocity of the advancing front and $\sqrt{ge}$ is the velocity of capillary waves in shallow water.

The condition $F_r > 1$ implies that the ridge front is a shock wave in shallow water, similar to the hydraulic jump commonly observed at the bottom of a kitchen sink (Figure 7.15).[33]

If we assume that the ridge has a constant height h between the radius R of the dry area and the radius R_F of the upstream front, we can write the standard Rankine-Hugoniot shock conditions:

- Conservation of mass:

$$h(V_F^2 - V^2) = eV_F^2 \tag{7.52}$$

FIGURE 7.15. Hydraulic jump observed upon impact of a water stream on a planar surface. A small defect placed within the jump zone triggers a shock wave, demonstrating that this flow zone is supercritical ($F_r > 1$). (From *Modeling Axisymmetric Flows* by S. Middleman. Academic Press (1995); Reproduced by permission.)

- *Conservation of momentum:*

$$eV_FV_1 = \frac{1}{2}g(h^2 - e^2) \tag{7.53}$$

where $V_1 = VR/R_F$ is the velocity of the liquid in the ridge at radius R_F. Note that energy is not conserved; a shock implies some dissipation.

These last two equations can be solved for V and V_F. The result is

$$V^2 = \frac{1}{2}g\frac{h^2 - e^2}{e} \tag{7.54}$$

and

$$V_F^2 = \frac{1}{2}g\frac{h(h - e)}{e}. \tag{7.55}$$

If we assume $h = e_c$ in equation (7.54), we recover Culick's law.

Both laws are verified experimentally with $h = 2.7$ mm, a value that is quite different from e_c. The discrepancy suggests that part of the surface energy is lost via other (unspecified) mechanisms.

Using the laser beam deflection technique, A. Buguin was able to measure the film's profile during the dewetting process (Figure 7.16) in the inertial regime ($R_e \gg 1$), both in the presence and in the absence of shock waves.

As the film's thickness e increases, the height of the ridge approaches $e'_c = 2.7$ mm, in agreement with the measurement deduced from the velocity [equation (7.54)]. The telltale sign of shock waves is the appearance of capillary waves associated with the moving ridge.

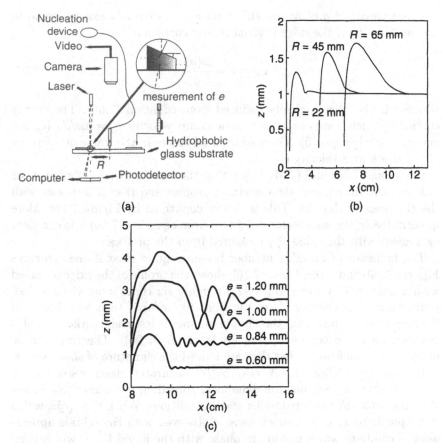

(a)

(b)

(c)

FIGURE 7.16. (a) Laser beam deflection technique used to measure the profile $z(x)$ of the film during wetting; (b) profile in the regime $F_r < 1$, when shock waves and ripples are absent; (c) profile for four different thicknesses 8 cm away from the nucleation point in the regime $F_r > 1$; the ripples are the signature of shock waves (courtesy A. Buguin).

In "shallow water," the dispersion relation for waves of small amplitude is given by: *

$$\omega = \sqrt{geq}\left(1 + \frac{q^2}{\kappa^2}\right)^{1/2} \tag{7.56}$$

where κ^{-1} is the capillary length.

---

*This result can be derived by writing that the sum of the surface energy and the grav-itational energy must equal the kinetic energy. For a film thickness $e = e_0 + ue^{iqx}e^{i\omega t}$ with an amplitude fluctuation u, we have: $F_q = (1/2)(\gamma q^2 + \rho g)u^2 = (1/2)\rho eV^2$, where V is related to $\dot{u} = du/dt$ via the relation $qV \propto \dot{u}/e$ (conservation of volume).

The wavevector q of waves with a phase velocity ω/q exactly equal to the velocity V_F of the ridge is given by the condition[30]

$$V_F = c = \frac{\omega(q)}{q} \qquad (7.57)$$

where c is the phase velocity deduced from equation (7.56). The energy carried by such waves progresses at a group velocity $c_g = d\omega/dq$ ($c_g > c$ and $c_g \to c$ when $q \to 0$). The wavefront advances in the frame of reference of the shock at a velocity $c_g - c$.

Equations (7.56) and (7.57) imply that the wavelength $\lambda = 2\pi/q$ is time independent throughout the dewetting process and that λ increases with the thickness of the film. This is clearly confirmed by Figure 7.16c. More quantitatively, the wavevector q deduced from equation (7.56) is in excellent agreement with the value of q measured from the profiles.

The influence of Froude's number becomes quite clear if one compares Figures 7.16b and 7.16c. Figure 7.16b shows the profile of the ridge obtained with a mixture of water and glycerol roughly six times more viscous than pure water. The velocity V_F of the front (6 cm/s) is then less than $\sqrt{ge}$ ($= 10$ cm/s), meaning that the Froude number is less than 1 [the Reynolds number, on the other hand, remains large ($R_e \approx 20$)]. The ripples have disappeared, confirming that they are indeed the signature of shock waves.

In summary, Culick's law $V = \sqrt{|S_d|/e\rho}$ accurately describes dewetting in the inertial regime, but the dynamical spreading parameter $|S_d|$ is less than the value $|S|$ measured under static conditions. When $V > \sqrt{ge}$, which corresponds to $F_r > 1$, a shock wave is observed with the telltale appearance of capillary waves moving in phase with the liquid front, which itself advances at a velocity V_F on a film at rest ($V_F = c$).

7.3.3   Liquid/Liquid Inertial Dewetting

The previous discussion can be extended to the case of a liquid film on a liquid substrate. X. Noblin has studied the inertial dewetting of a film of water deposited on carbon tetrachloride, which is a highly hydrophobic surface, as evidenced by the large thickness of the puddles $e_c = 7$ mm.[34] The dewetting velocity is extremely high, of the order of a m/s for thin films. Since a liquid substrate constitutes an ideal surface, one can create much thinner metastable films than is possible on a solid and, therefore, attain larger velocities. Under such conditions, Culick's law is perfectly verified.

However, the ridge can become coupled to the capillary waves in a deep bath characterized by a surface tension $\gamma = \tilde{\gamma}_B = \gamma_{AB} + \gamma_A + (1/2)\rho g e^2$ upstream of the ridge, and $\gamma = \gamma_B$ downstream, where the film has dried up.

When the thickness was decreased, a whole sequence of hydraulic shocks was observed during dewetting (see Figure 7.17).

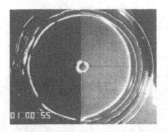

FIGURE 7.17. Inertial dewetting of water floating on a pool of carbon tetrachloride (CCl$_4$) for three different thicknesses. Left picture: $e = 0.73$ mm, with no ripples and no shock; middle picture: $e = 0.61$ mm, with ripples toward center and shock waves directed downstream; right picture: $e = 0.43$ mm, with ripples toward the center and the outside, shock waves upstream and downstream (courtesy X. Noblin).

7.4 Visco-Elastic Dewetting

We are interested now in the opposite limit, namely, the dewetting of ultra-viscous "pastes," with a viscosity roughly a million times greater than that of water, deposited on a low-viscosity liquid or simply suspended in air.

For the conditions described in this section to apply, the surface viscosity $\eta_S = \eta e$ of the film must be greater than the viscosity of the substrate multiplied by the size of the hole at all times during the dewetting process. If this criterion is not met, the dissipation in the substrate dominates and the opening of the hole (discussed in section 7.2.3) becomes independent of the viscosity of the film.

Investigations of the rupture of "bare" films (i.e., without added surfactants) are quite recent.[35, 36] By contrast, the bursting of soap films has been studied for well over 100 years, notably by K. Mysels.[37, 38] The bursting is inertial in that the surface energy is converted into kinetic energy, with exactly half the total energy being dissipated in a form that remains to be elucidated. A detailed discussion of soap films will be given in section 8.4.4.

When the film is ultra-thin, viscous dissipation becomes dominant. For "bare" films, the viscous regime should prevail below a thickness e_0 [29]:

$$e_0 \approx \frac{\eta^2}{\gamma \rho}. \tag{7.58}$$

For soapy water, $e_0 \approx 10$ nm. For a paste, on the other hand, e_0 is measured in kilometers, and we can rest assured that we are then always in the viscous regime.

Applications of bursting viscous films are numerous. They include

- *the life of viscous foams without surfactants:* polymer foams, foams in the food industry, mineral glass foams, mineral "foams" formed during volcanic eruptions (Figure 7.18a), mud bubbles formed by the release of gas (Figure 7.18b),

(a)   (b)   (c)

FIGURE 7.18. A few more exotic forms of bubbles and foams: (a) lava bubble (mineral)—Courtesy Hoa-Qui, Bourseiller, and Durieux. (b) mud bubble (viscous foam)—Courtesy Hoa-Qui and Krafft. (c) vesicles (courtesy P.-H. Puech and L. Moreaux).

- *coalescence of emulsions:* ultra-viscous oil in water,
- *controlled opening of transient pores in vesicles* (Figure 7.18c): microcapsules of surfactants used to simulate biological membranes or to deliver drugs in the human body.[39,40]

We begin by discussing the laws of bursting of an ultra-viscous polymer film (PDMS) freely suspended or deposited on a low-viscosity fluorinated oil.[33] Next, we will show that bursting in molten glass behaves in a similar way. Finally, we will close this chapter by studying the draining and bursting of viscous bubbles.[35]

7.4.1 Rupture of Ultra-Viscous Films

We consider films that are either suspended in air or deposited on a low-viscosity substrate. The only difference between these two situations is the nature of the driving force F_M responsible for opening up the hole. $F_M = 2\gamma$ for a suspended film, whereas $F_M = |S|$ for a film on a liquid substrate, the only purpose of which is to control the wettability and to provide an opportunity to create thicker films.

G. Debregeas has conducted studies of ultra-viscous suspended PDMS films, which he made by dipping a ring in a solution of polymer (Figure 7.19a). As the solvent evaporated, it left behind a film of thickness e ranging from a few microns to about 100 μm, as measured by infrared spectroscopy. At about the same time, P. Martin was studying holes developing in thicker PDMS films ($e \approx 0.1$ to 1 mm) floating on low-viscosity PFAS. In both cases, the bursting process was recorded with a video camera and the thickness of the film was monitored at every point by means of an optical interference technique.

The laws governing the dynamics of the process proved to be entirely different.

1. The hole opens up according to an exponential law (Figure 7.19b) given by

$$R = R_0 \exp\left(\frac{t}{\tau}\right) \tag{7.59}$$

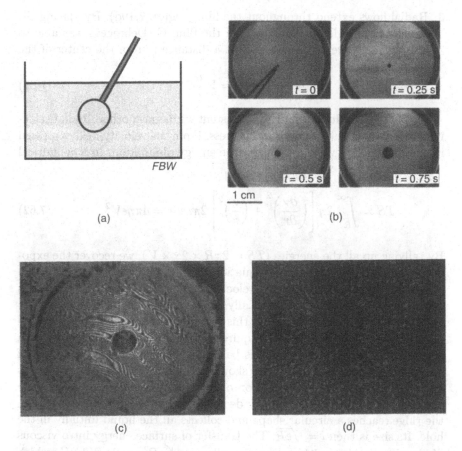

FIGURE 7.19. A hole in the process of opening up in a bare film of ultra-viscous PDMS. (a) formation of the film; (b) the opening of a hole is initiated by touching the film with a needle; (c) interference pattern showing the absence of a ridge; (d) velocity field visualized by the displacement of microbeads (courtesy G. Debregeas).

where $\tau = \eta e / \gamma$. In contrast to all the cases described previously, where the dewetting velocity was constant, here the velocity increases proportionately to the radius of the hole:

$$V = \dot{R} = V^* \frac{R}{e} \qquad (7.60)$$

where $V^* = \gamma / \eta$. The velocity V^* is quite small (~ 20 µm/s), but the factor R/e itself can become extremely large for very thin films ($R = 1$ cm and $e = 1$ µm gives $R/e = 10^4$).

2. There no longer is a ridge. Interferograms (Figure 7.19c) show that the film thickness increases uniformly and no liquid accumulates anywhere around the hole.

3. Radial flows extend throughout the film (Figure 7.19d). By placing silvered microbeads on the surface of the film, G. Debregeas was able to determine the velocity field $v(r)$ at a distance r from the center of the hole. He found

$$v(r) = V\frac{R}{r}. \qquad (7.61)$$

These three results are entirely consistent with one another. Radial velocities indeed do lead to a uniform thickness. From the velocity field expressed in equation (7.61), we can calculate the energy dissipation. In a cylindrical coordinate system, we get

$$T\dot{S} = \int_{R}^{\infty} 2\eta \left[\left(\frac{\partial v}{\partial r}\right)^2 + \left(\frac{v}{r}\right)^2 \right] 2\pi r e\, dr = 4\pi \eta e V^2. \qquad (7.62)$$

By tallying up all the energies ($T\dot{S} = 2\pi R \times 2\gamma \times V$), we recover the exponential growth law contained in equations (7.59) and (7.60) since $\dot{R} = \frac{\gamma}{\eta e}R$.

We should point out that the velocity at which the hole opens up does not keep on increasing exponentially forever. Eventually, it is bound to reach some upper limit, although this regime has never been observed experimentally because it would require huge liquid surface areas. From a theoretical standpoint, an analysis based on the viscoelastic properties of the polymer predicts that a ridge should form after sufficiently long times and that the increase in dissipation ought to slow the process down. The analysis suggests that the velocity decreases toward a limiting value when the ridge reaches a circular shape and collects all the liquid initially in the hole. Its size is then $l = \sqrt{e R}$. The transfer of surface energy intro viscous dissipation can be written dimensionally as $\eta \frac{V^2}{l^2} R^2 e = 2\pi R(2\gamma)V$, which amounts to $V \propto \gamma/\eta$.

Yet another hydrodynamic model of the bursting of a purely viscous film leads to a similar exponential growth, although for long times the velocity keeps on increasing up to the limit of inertial bursting $V = \sqrt{2\gamma/\rho e}$ beyond a radius $R_c = \eta \left/ \sqrt{\frac{e}{\gamma\rho}} \right.$.[39]

The present study has also been extended to the bursting of molten glass. Near the glass transition temperature, molten glass has a high viscosity (of the order of 10^6 mPa-s), while its viscoelastic properties elastic are vastly different because the elastic modulus of glass is extremely high (10^4 times greater than that of molten polymer), and the relaxation times are correspondingly much shorter. G. Debregeas placed a thin glass plate (with a hole pierced through its center) in an oven and made a video recording of the subsequent expansion of the hole. The growth turned out again to be exponential in time. After allowing the plate to cool, its examination showed that the film had grown uniformly without any ridge forming on the periphery of the hole. This experiment makes it possible to estimate the surface tension of molten glass as long as the viscosity $\eta(T)$ is known as a function of temperature. It is useful for measuring the lifetime of glass foams, which oftentimes prove to be a nuisance when they develop in ovens.

Figure 7.18a shows another type of mineral foam formed by lava during volcanic eruptions. Very much the same morphology can be achieved on a smaller scale in the microwave oven you keep in your kitchen. Place about 20 sugar cubes in a bowl with a little water. Heat it for about one minute at high power and you will get a viscous sugar syrup. But if you inadvertently leave it in for 20 minutes, you will end up with a spectacular chunk of black carbon foam.

In general, foams may appear even in the absence of surfactants when we inject air in a liquid paste. They play an important role in various areas of organic and mineral chemistry. To better understand them, we proceed next to examine viscous bubbles.

7.4.2 Life and Death of Viscous Bubbles

In this section, we describe the bursting of viscous bubbles prepared by injecting air into liquid pastes (the bubble gum phenomenon).

Figure 7.20 shows a viscous bubble obtained by injecting a large air bubble in ultra-viscous PDMS. The bubble rises slowly in the liquid and emerges at the surface as a hemisphere (roughly 1 cm in size in the present case). Gravitational draining thins the bubble and the thickness $e(t)$ can be tracked as a function of time by optical interferometry. The draining laws turn out to be quite different from those governing soap bubbles.[33] The reason is that here the motion of the fluid is of the plug-flow type (the velocity in a slice of liquid is uniform). By contrast, in a soap bubble, water must flow between two surfactant layers and its velocity is zero at the boundaries of the film (giving a parabolic velocity profile of the Poiseuille type).

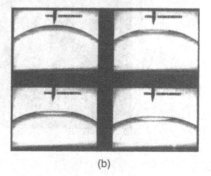

(a)   (b)

FIGURE 7.20. (a) A viscous bubble is formed by injecting air in ultra-viscous PDMS; it thins down by way of a draining process; (b) bursting and death of the bubble recorded with an ultra-high-speed camera. (From G. Debregeas, P. G. deGennes, and F. Brochard, in *Science, 279*, p. 1704 (1998). © 2001 American Association for the Advancement of Science. Reproduced by permission.)

With this optical technique, we can follow the thinning down of the bubble and create ultra-thin films (about 100 nm). The bubble can be made to burst by touching it with a needle. The bursting lasts about 1 ms, and an ultra-high-speed camera is needed to record the event. The expansion of the hole proceeds exponentially in time with a velocity $V = \gamma R/\eta e$ (about 1 m/s)—of the same order of magnitude as the bursting velocity of a soap bubble. When the hole is large, the air under pressure that once supported the film escapes rapidly and the bubble settles somewhat like a parachute upon reaching the ground.

Quite recently, the study of viscous bubbles has proved useful in interpreting the formation of transient pores in objects whose size is 10,000 times smaller. We are referring to giant vesicles of phospholipids, which are amphiphile molecules hooking together to form membranes.[40,41] These giant vesicles, roughly 100 μm in size, are artificial objects mimicking a red globule in that a lipid bilayer encapsulates some internal medium. At equilibrium, the vesicles have zero surface tension because they constitute a "closed" system that optimizes its surface energy. That is the reason they exhibit phenomenally large fluctuations (Figure 7.21a), with amplitudes

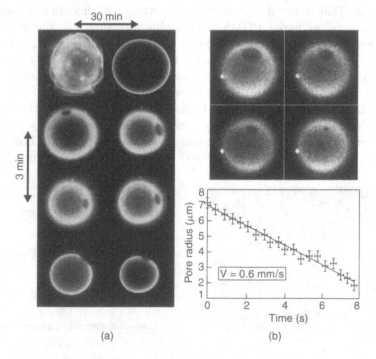

(a)   (b)

FIGURE 7.21. (a) When subjected to an intense light for a period of 30 min, a giant vesicle is placed under mechanical stress. Transient pores open up, close down, and are reborn every 3 min. (b) Closure of a hole in a vesicle. (From "Dynamics of Transient Pores in Stretched Vesicles." by O. Sandre, L. Moreaux, and F. Brochard. In *Proceedings of the National Academy of Sciences.* © 2001 National Academy of Sciences, USA. Reproduced by permission.)

1,000 times larger than those experienced by a soap bubble. Such huge fluctuations are possible because they are limited by very weak curvature energies of the membrane. The same property also happens to be responsible for the phenomenon of red globule scintillation, which manifests itself as a modulation of transmitted light because of the encapsulated hemoglobin solution.[42] This lack of tension makes vesicles extremely robust. Instead of expanding, a hole pierced in their distended membrane actually closes up in order to minimize its border energy. As a result, these objects do not burst like fragile soap bubbles. On the other hand, it is possible to inflate these giant vesicles and place them under mechanical tension. Figure 7.21a shows what happens to an initially fluctuating vesicle when it is being inflated by exposure to an intense light.* The observations were done by fluorescent microscopy. The tensions σ involved are quite weak (of the order of 10^{-2} mN/m), which is a hundred to a thousand times less than the surface tension of viscous bubbles. Beyond that, the membranes disintegrate. Do such stretched vesicles become as fragile as soap bubbles? Figure 21a shows that, instead of bursting, the vesicle gives birth to a succession of transient pores. Figure 7.21b shows the life cycle of a pore. A hole opens up and expands up to a maximum radius r_M, after which it closes up again! In other words, the bursting is not irreversible. Instead, the pore closes up slowly (over a period of several seconds). Pores open up because the vesicle is under tension. However, in contrast to viscous bubbles, the tension σ of the membrane is not constant. Rather, it relaxes via two mechanisms:[41]

1. The hole opens up. This relaxes the membrane by shrinking its surface area.
2. The internal fluid escapes. As soon as a pore is formed, the Laplace pressure sets up a liquid "jet," which deflates the vesicle.

Finally, the pore closes up completely so as to minimize the border energy of the hole. The cycle can then start over again if the system is still under illumination.

In this chapter, we have described the various processes by which liquid films can recede either by "dewetting" or by bursting when suspended or floating on a liquid. The reader may complement his/her understanding of the topic by reading a number of pertinent popular articles.[42]

References

[1] A. Sharma and E. Ruckenstein, *J. Colloid Interface Sci.* **106**, 12 (1985); J. E. Proust, S. D. Tchaliovska, and L. Ter-Minassian-Saraga, *J. Colloid Interface Sci.* **98**, 319 (1984).

[2] G. I. Taylor and E. Michael, *J. Fluid Mech.* **58**, 625 (1973); D. F. James, *J. Fluid Mech.* **63**, 657 (1974); A. Sharma and E. Ruckenstein, *J. of*

---

*How light creates surface tension is not understood at present.

188   References

Colloid Interface Sci. **137**, 443 (1990); C. Sykes, *C. R. Acad. Sci. Paris* **313**, 607 (1991); A. Sharma, *J. of Colloid Interface Sci.* **156**, 96 (1992).

[3] J. F. Padday, *Spec. Discuss. Faraday Soc.* **1**, 64 (1971); C. Sykes, C. Andrieu, V. Détappe, and S. Deniau, *J. Phys. III France* **4**, 775 (1994).

[4] G. Debregeas and F. Brochard-Wyart, *J. Colloid Interface Sci.* **190**, 134 (1997).

[5] C. Redon, F. Brochard-Wyart, and F. Rondelez, *Phys. Rev. Lett.* **66**, 715 (1991).

[6] G. Reiter, *Phys. Rev. Lett.* **68**, 75 (1992); *Langmuir* **9**, 1344 (1994); *Macromolecules* **27**, 3046 (1994).

[7] G. Reiter, R. Khanna, and H. Sharma, *Phys. Rev. Lett.* **85**, 1432 (2000).

[8] R. Khanna, H. Sharma, and G. Reiter, *Eur. Phys. Journal* **2**, 1 (2000).

[9] J. B. Broszka, N. Shahizadeh, and F. Rondelez, *Nature* **360**, 24 (1992).

[10] J. Sagiv, *J. Am. Chem. Soc.* **102**, 92 (1980).

[11] P. G. de Gennes, *C. R. Acad. Sci. (Paris)* **303**, 1275 (1986).

[12] F. Brochard-Wyart, J. M. di Meglio, and D. Quéré, *C. R. Acad. Sci. (Paris)* **304**, 553 (1987).

[13] F. Brochard-Wyart and P. G. de Gennes, *Adv. Colloid Interface Sci.* **39**, 1 (1992).

[14] C. Andrieu, C. Sykes, and F. Brochard-Wyart, *J. Adhesion* **58**, 15 (1996).

[15] F. Brochard-Wyart, C. Redon, and F. Rondelez, *C. R. Acad. Sci. (Paris)* **306**, 1143 (1988).

[16] C. Andrieu, C. Sykes, and F. Brochard-Wyart, *Langmuir* **10**, 2077 (1994); C. Andrieu, *Doctoral thesis, University of Paris* (1995).

[17] L. Bacri, *Doctoral thesis, University of Paris* (2000); L. Bacri and F. Brochard-Wyart, *Europhys. Lett.* **56**, 414 (2001).

[18] C. Redon, J. B. Brzoska, and F. Brochard-Wyart, *Macromolecules* **27**, 468 (1994).

[19] P. G. de Gennes, *C. R. Acad. Sci. (Paris)* **288**, 219 (1979).

[20] G. Reiter and R. Khanna, *Phys. Rev. Lett.* **85**, 2753 (2000).

[21] P. Martin, A. Buguin, and F. Brochard-Wyart, *Europhys. Lett.* **28**, 421 (1994).

[22] J. Haffel and H. Brenner, *Low Reynolds Number Hydrodynamics*, (Englewood Cliffs, N.J.: Prentice-Hall, 1965), chap. 7, section 7.

[23] D. Andelman, J. F. Joanny, and M. O. Robbins, *Europhys. Lett.* **7**, 731 (1988).

[24] A. R. Karim, J. F. Douglas, C. C. Han, and R. A. Weiss, *Phys. Rev. Lett.* **81**, 1251 (1998).

[25] G. Reiter, *Science* **282**, 888 (1998).

[26] S. Herminghaus, et al, *Science* **282**, 916 (1998); S. Seemann, S. Herminghaus, and K. Jacobs, *J. Phys. Condensed Matter* **13**, 1 (2001).

[27] O. Ou Ramdane, *Doctoral thesis, University of Paris* (1998).

[28] F. Brochard-Wyart, P. Martin, and C. Redon, *Langmuir* **9**, 3682 (1993).

[29] F. Brochard-Wyart, E. Raphaël, and L. Vovelle, *C. R. Acad. Sci. (Paris)* **321**, 367 (1995).

[30] A. Buguin, L. Vovelle, and F. Brochard-Wyart, *Phys. Rev. Lett.* **83**, 1183 (1998).

[31] F. E. Culick, *J. Appl. Phys.* **31**, 1128 (1960).

[32] Lord Rayleigh, *Proc. Roy. Inst.* **13**, 261 (1891).

[33] Stanley Middleman, *Modeling Axisymmetric Flows: Dynamics of Films, Jets, and Drops* (San Diego, CA: Academic Press, 1995).

[34] X. Noblin, *DEA Internship Report, University of Paris* (2000).

[35] G. Debregeas, P. Martin, and F. Brochard-Wyart, *Phys. Rev. Lett.* **75**, 3686 (1995).

[36] G. Debregeas, P. G. de Gennes, and F. Brochard-Wyart, *Science* **279**, 1704 (1998); G. Debregeas, *Doctoral thesis, University of Paris* (1997).

[37] W. E. Ranz, *J. Appl. Phys.* **31**, 1128 (1960)

[38] S. Frankel and K. J. Mysels, *J. Phys. Chem.* **73**, 3028 (1969).

[39] M. Brenner, D. Gueyffier, *Phys. Fluids* **11**, 737 (1999).

[40] O. Sandre, L. Moreaux, and F. Brochard-Wyart, *Proc. Natl. Acad. Sci.* **96**, 10591 (2000); O. Sandre, *Doctoral thesis, University of Paris* (2000).

[41] F. Brochard-Wyart, P. G. de Gennes, and O. Sandre, *Physica A* **278**, 32 (2000).

[42] F. Brochard-Wyart, Une bulle dégonflée: le globule rouge (A Deflated Bubble: The Red Globule), *La Recherche* **75**,173 (1997); P. Martin, G. Debregeas, and F. Brochard-Wyart, La dynamique des bulles perforées (The Dynamics of Perforated Bubbles), *Pour la Science* (May 1995); F. Brochard-Wyart, Eclatement des films: de l'Hélium superfluide aux pâtes visqueuses (Bursting Films: From Superfluid Helium to Viscous

Pastes), F. Brochard-Wyart, *Bulletin de la SFP* 1032 (March 1996); F. Brochard-Wyart, Quand les liquides démouillent: des surfaces sèches, archisèches, qui ont séché sans sécher (When Liquids Dewet: Dry and Ultra-Dry Surfaces that Dried off without Really Drying), *La Recherche* (April 1996).

8
Surfactants

8.1 Frustrated Pairs

8.1.1 Principle

When we wash our dishes after dinner, we add to warm water a few drops of a miracle product—soap, or better yet, a dishwashing liquid detergent. Instantly, the water starts foaming. Why is that?

The miracle product is what we call a *surfactant*. Its molecule has two parts with different affinities. A "hydrophilic" part, which is eager to mix with water, and a "hydrophobic" part, which abhors water. Figure 8.1 shows three major examples of surfactants.

In most cases, the hydrophobic part is formed by one (or more) aliphatic chain $CH_3(CH_2)_n$. The hydrophilic part can be an ion (either anion or cation), which forms a "polar head." The head likes liquids with a high dielectric constant such as water. In other cases, the hydrophilic part can be a short chain of neutral units that are soluble in water. An example is POE (polyethylene oxide), formed of units of the type (CH_2-CH_2-O). Typically, there are a dozen or so carbon atoms in the aliphatic part, and five or six ethylene oxide groups in the hydrophilic part. Table 8.1 gives a somewhat more complete view of the hydrophilic and hydrophobic groups that are currently in use.

What we have just described is reminiscent of a frustrated couple, where both partners are bound together but have very different tastes. How will such couples behave, either individually or in large numbers? The question is just as relevant in sociology as it is in the physical chemistry of surfaces. We proceed next to analyze a few responses at the molecular level.

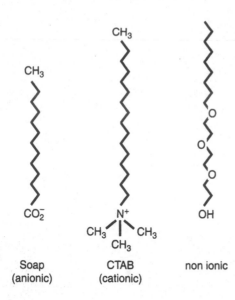

FIGURE 8.1. Three examples of surfactants. The middle one is CTAB (cetyl trimethyl ammonium bromide).

Soap
(anionic)

CTAB
(cationic)

non ionic

TABLE 8.1. Usual constituents of surfactants.

Hydrophobic part		Hydrophilic part	
Formula	Name	Formula	Name
$CH_3-(CH_2)_n-$	Alkyl	$-O-SO_3{}^-Na^+$	Sulfate
$CH_3-(CH-CH_2)_n$ $\vert$ CH_3	Branched alkyl	$-CO-O^-Na^+$	Carboxylate
$CH_3-(CH_2)_n-$phenyl	Alkyl phenyl	$-N^+(CH_3)_3Cl^-$	Quaternary ammonium
$CF_3-(CF_2)_n-$	Perfluoroalkyl	$-(O-CH_2-CH_2)_nOH$	Polyethoxy

8.1.2   The Notion of Hydrophilic/Lipophilic Balance (HLB)

The first question to examine is whether one of the two partners is dominant. In the present context, this amounts to asking whether the surfactant is more soluble in water or in oil (the simplest example of oil is a saturated hydrocarbon). Assume, for instance, that we deposit some oil on top of water in a test tube, and that we then add a small quantity of a surfactant. At equilibrium, a constant ratio is established between the concentration of the surfactant in oil and its concentration c_w in water:*

$$\frac{c_0}{c_w} = k. \tag{8.1}$$

---

*Some surfactant also exists at the boundary between water and oil, but the corresponding mass is negligible in comparison to the mass throughout the volume.

The most natural classification scheme for surfactants would then be based on the value of the segregation coefficient k. The surfactants for which $k > 1$ are those that have a long aliphatic chain, whereas those for which $k < 1$ have a short one.

Although a classification method based on k is logical, it suffers from a number of practical shortcomings.

- Which particular oil is chosen has an influence.
- Oftentimes, one encounters extreme cases in which k is either exceedingly large (surfactants insoluble in water) or exceedingly small (surfactants insoluble in oil). In either case, measuring k becomes all but impossible.

For these reasons, it has become customary to use a somewhat different definition introduced in 1946 by Griffin. It is based on a simple-minded estimate of the work Δ involved in transferring a molecule from water to oil. In general, we must have

$$k = \exp(-\Delta/kT). \tag{8.2}$$

The value of Δ is estimated by means of the expression

$$\Delta = U_H n_H - U_L n_L \tag{8.3}$$

where U_H is the transfer energy of one hydrophilic group and n_H is the number of such groups in a molecule. Likewise, U_L and n_L are the corresponding numbers for the "lipophilic" part (which likes oil). The quantities U_L and n_L can be determined empirically in a few cases. In this approximation, the base-10 logarithm of the segregation coefficient is of the form

$$\log_{10}(k) = m_L n_L - m_H n_H \tag{8.4}$$

where m_L and m_H are phenomenological coefficients.

What has come to be known as the HLB *(hydrophilic lipophilic balance)* is essentially proportional to $(-\log_{10} k)$, albeit with minor modifications. The neutral case $(k = 1)$ is arbitrarily assigned an HLB value of 7 by analogy with the pH values of acids and bases in water. This convention leads to

$$\text{HLB} = 7 + \alpha(m_H n_H - m_L n_L) = 7 + (\tilde{m}_H n_H - \tilde{m}_L n_L) \tag{8.5}$$

where $\tilde{m} = \alpha m$. The normalization constant α in this formula was chosen to make the HLB values of usual surfactants fall in the range 0 to 14, just as ordinary pH values do. Table 8.2 lists values of the parameter $\tilde{m}$ for a few important chemical functions.

The present definition of HLB is far from perfect. It assumes, for instance, that when a surfactant molecule is immersed in water, all the hydrocarbon segments are exposed to water. If so, the number of contacts should be n_L, and the resulting energy ought to be $n_L U_L$. In reality, the hydrocarbon

TABLE 8.2. HLB coefficients.

Hydrophilic groups	$\tilde{m}_H$	Hydrophobic groups	$\tilde{m}_L$
—CO_2Na	19.1	—CH_2	0.47
—SO_3Na	11.0		
—$N-(CH_3)_3$–Cl	9.4		
—O–	1.3		
—OH	1.9		

TABLE 8.3. Practical properties associated with HLB ranges.

HLB	Applications
1.5–3	Antifoaming agents
3–6	Water/oil emulsions
7–9	Foaming agents Wetting agents
8–18	Oil/water emulsions
13–15	Detergents
15–20	Dissolving agents for organic products

part tends to fold back on itself in an attempt to diminish such unfavorable contacts, and the energy ends up being overestimated.

For reasons of this type, numerous authors have proposed improved scales of HLB values. But given the introductory level of the present book, we shall ignore these refinements. Our aim is merely to explain (and gradually at that) some of the empirical data appearing in Table 8.3.

8.2   Aggregation of Surfactants

8.2.1   Aggregation in Volume: Micelles

We take the example of an ionic surfactant. When placed in water, a group consisting of a number p of surfactant molecules can associate to form a spherical structure illustrated in Figure 8.2, where all the polar heads face water and all the aliphatic tails are protected from it. Such an aggregate, or *micelle*, was conceived of by McBain in 1913.

Its diameter $2R$ depends on several factors:

- *The area Σ_p per polar head.* Most often, interactions between neighboring molecules impose a fairly well defined value of Σ_P.
- *The volume V per chain.* If we neglect the head (on the grounds that it is much smaller than the tail) we simply have $V = n_L v_L$, where v_L is the volume per aliphatic group.

FIGURE 8.2. "Steric" model of a spherical
micelle in water. Each surfactant molecule
has a tail of length R and a polar head
whose surface area exposed to water has a
prescribed value Σ_p.

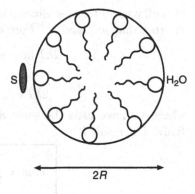

At this point, the geometry of the micelle, which has a radius R, is fully
determined. Its volume is

$$\frac{4\pi}{3}R^3 = pV \cong pn_L v_L. \tag{8.6}$$

Its surface area is

$$4\pi R^2 = p\Sigma_p. \tag{8.7}$$

From this we deduce the value of the radius:

$$R = 3n_L \frac{v_L}{\Sigma_p}. \tag{8.8}$$

Finally, the number p of molecules per micelle is given by

$$p = 36\pi \frac{v_L^2 n_L^2}{\Sigma_p^3} \approx 100 \frac{v_L^2 n_L^2}{\Sigma_p^3} \quad \left(\approx n_L^2\right). \tag{8.9}$$

In practice, the number p is of the order of 100.

Micelles form only above a certain *critical micellar concentration* (commonly abbreviated as *cmc*). The *cmc* can be estimated by means of a crude
thermodynamic argument based on chemical potentials.

Consider the case of an ionic surfactant. The chemical potential of the
unassociated surfactant $\mu^{(1)}$ is of the form[1]

$$\mu^{(1)} = \mu_0^{(1)} + 2kT \ln(cV). \tag{8.10}$$

The factor of 2 comes about because we have chosen an electrically charged
surfactant. Therefore, to each surfactant ion there corresponds a matching
mobile counter-ion.

The chemical potential $\mu^{(M)}$ of a molecule tied up in a micelle has a
similar form:

$$\mu^{(M)} = \mu_0^{(M)} + \frac{kT}{p} \ln\left(\frac{cV}{p}\right). \tag{8.11}$$

Here the entropic term (proportional to kT) is divided by p because the
translational degrees of freedom are shared by the p molecules of a micelle

plus all its counter ions. Since p is large, this term can actually be neglected. Writing that $\mu^{(1)}$ and $\mu^{(M)}$ are equal leads to an equation for the *cmc*:

$$2kT \ln(cmc \cdot V) = \mu_0^{(M)} - \mu_0^{(1)} \tag{8.12}$$

or

$$2kT \ln(cmc \cdot V) = -n_L U_L \tag{8.13}$$

where we have used the same approximation as in our discussion of the HLB. The result is

$$cmc = \frac{1}{V} \exp\left(-\frac{n_L U_L}{2kT}\right) \tag{8.14}$$

Its is clear that if n_L is large (long chains), the *cmc* can be quite small.

For concentrations $c < cmc$, the surfactant exists in the form of isolated molecules. For concentrations $c > cmc$, the majority of the surfactant finds itself aggregated in micelles. A fraction of the surfactant, however, does remain in a non-aggregated form. Its concentration is constant and equal to the *cmc* so as to maintain the micelle/isolated molecule equilibrium described by equation (8.12).

The previous discussion is grossly simplified. There are some surfactants, notably those that have two chains per head, which do not form micelles at all. Furthermore, at higher concentrations of conventional surfactants, it is possible to create cylinders and bilayers, rather than simple spherical micelles. Still, for most ordinary applications of surfactants, the normal trend is to avoid such high concentrations, which generate highly viscous systems that are more difficult to handle.

8.2.2 Water/Air Interfaces

We now return to our kitchen sink filled with warm water, to which we add a few drops of a surfactant. When we insert a frustrated pair in water, the polar head is happy, whereas the aliphatic chain is miserable. One way to improve this state of affairs is to place the molecule at the water/air interface (Figure 8.3). This strategy allows us to realize what is referred to as a *monolayer* of surface density Σ^{-1} (where Σ is the surface area per polar head).

A monolayer can be created in one of two ways:

1. It can be made from a solution (as we have just described), provided that the surfactant is sufficiently soluble (which implies an aliphatic chain that is not too long).
2. In the opposite case, it is possible to deposit the surfactant with a pipette on the water surface and see it spread as a monolayer.

We will first discuss the case of insoluble monolayers, which were first studied by Langmuir.

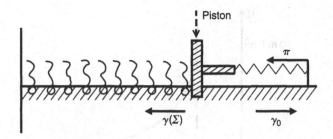

FIGURE 8.3. Langmuir's setup: Compression of an insoluble monolayer.

8.2.2.1 Insoluble Monolayers

Langmuir devised an experimental setup that allowed him to compress
the layer, in the manner illustrated in Figure 8.3. Since the surfactant is
insoluble, the compression takes place at a constant number of molecules.
A piston provides the means to apply a *surface pressure* $\Pi(\Sigma)$ on the layer.
This pressure is related to the free energy $F(\Sigma)$ per molecule via the usual
relation:

$$\Pi(\Sigma) = -\left.\frac{\partial F}{\partial \Sigma}\right|_T. \tag{8.15}$$

The pressure Π is also related to the surface tension $\gamma(\Sigma)$ in the presence
of surfactants. We can see in Figure 8.3 that the piston is in equilibrium
under the simultaneous action of the following forces:

1. $\gamma(\Sigma)$, pulling inwards,
2. γ_0, the surface tension of pure water, pushing outwards,
3. the force Π, pulling inwards.

   Therefore, we have $\gamma(\Sigma) + \Pi = \gamma_0$, which leads to the fundamental relation:

$$\gamma(\Sigma) = \gamma_0 - \Pi. \tag{8.16}$$

The qualitative shape of the $\Pi(\Sigma)$ curves is sketched in Figure 8.4. In
certain cases, there is clear evidence of a horizontal floor, indicating a liq-
uid/gas phase transition in two dimensions, which occurs at a very low
pressure. In practice, however, one almost never operates on the gas side
of the graph. In the liquid phase and at high pressures, the surface area Σ
tends toward a limiting value Σ_0 corresponding to a highly compact state.

   The cascade of transitions that can occur in compression is often far
more complicated, but for our present purpose, the description we have
just given is adequate.

8.2.2.2 Soluble Monolayers

Here, if we conduct experiments under pressure, part of the monolayer can
redissolve in the underlying water. We can, however, monitor Σ (and Π) to
some extent by varying the concentration c of the surfactant in the water.

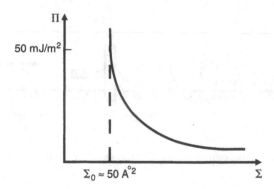

FIGURE 8.4. Equilibrium curve relating the surface area Σ per polar head to the applied surface pressure Π.

What is the relation between Σ and c? We can find the answer by considering the surface/volume equilibrium in terms of chemical potentials. In the bulk liquid (and as long as the concentration is below the cmc), the chemical potential of the surfactant was described by equation (8.10) if the surfactant is of the ionic type. Here we will consider a surfactant that is neutral, giving us the opportunity to illustrate a different case. In this instance, the chemical potential of the surfactant dissolved in the volume of the liquid is

$$\mu_{VOL} = \mu_W^0 + kT\ln(cV) \tag{8.17}$$

where, as before, V is the volume of a surfactant molecule. At the surface, we cannot make the assumption of a dilute solution. What we can do is derive μ from the thermodynamic relation

$$\mu_{SURF} = F(\Sigma) + \Sigma \cdot \Pi(\Sigma). \tag{8.18}$$

At constant temperature, this leads to

$$d\mu_{SURF} = \Sigma \cdot d\Pi. \tag{8.19}$$

This relation was derived early on by Gibbs. Gibbs was one of the founders of statistical physics; he was also a great devotee of soap films! From equation (8.19) we can obtain the chemical potential of the surfactant with the help of the curves describing $\Pi(\Sigma)$.

$$\mu_{SURF} = \mu_{SURF}^0 + \int_0^\Pi \Sigma \cdot d\Pi \tag{8.20}$$

where μ_{SURF}^0 represents the free energy of a single surfactant molecule at the surface. It is lower than μ_W^0 because such a molecule prefers to float between water and air, rather than to be immersed in the water bath.

Equating μ_{SURF} and μ_{VOL} leads to a relation between c and Σ:

$$kT\ln(cV) = \mu_{VOL} - \mu_{SURF}^0 - \int_0^\Pi \Sigma \cdot d\Pi. \tag{8.21}$$

FIGURE 8.5. Surface tension of a monolayer as a function of the concentration c
of the surfactant in solution.

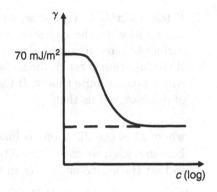

It is often convenient to recast this result in the form of Gibbs's differential
equation [equation (8.19)]:

$$kT\frac{dc}{c} = -\Sigma \cdot d\Pi = \Sigma \cdot d\gamma(\Sigma). \tag{8.22}$$

The surface tension $\gamma(\Sigma)$ can be measured directly as a function of the
concentration c. With the help of equation (8.22), we can derive the surface
concentration $1/\Sigma$ of the monolayer. A typical curve $\gamma(c)$ is schematically
depicted in Figure 8.5.

The figure shows that the surface tension decreases significantly by addition of the surfactant, by as much as a factor of 2 or 3. But it does not
decrease all the way to zero! There is a lower limit corresponding to the
cmc, at which point the chemical potential is pinned. When $c > cmc$, the
pressure Π and the surface area Σ remain constant.

Is it possible to reach $\gamma = 0$ by adding surfactants in carefully chosen
systems? The answer is no for air/water interfaces. The pressure would have
to increase up to $\Pi = \gamma_0$, in accordance with equation (8.16). Unfortunately,
for such high values of Π, the chemical potential μ_{SURF} invariably turns
out to be higher than $\mu_{MICELLE}$. In the competition between the surface
and micelles, the surface is almost always the loser because of the cost of
the interfacial energy between the aliphatic tails and air.

For an interface between oil and water, on the other hand, there exist some judiciously selected surfactants capable of achieving an interface
energy that is essentially zero. This phenomenon is responsible for the existence of *microemulsions*. They are systems of oil and water plus surfactants,
which spontaneously develop considerable surface areas.

8.2.2.3 Dynamical Surface Tensions

The discussion up to now was limited to *static* regimes. In some cases,
however, interfaces between water and air are being created rather quickly,
for instance, by injecting bubbles in a soap solution. The reader is referred
to the literature for details on this problem and its applications.[2]

1. If the injection is very slow, some of the surfactant can migrate from the solution to the interface, and experiments will measure the static surface tension discussed above.
2. If the injection is rapid, the surfactant has to diffuse toward the surface, which takes a finite time τ. If the distance to the interface is l, the laws of diffusion tell us that

$$l^2 = D\tau \tag{8.23}$$

where D is the diffusion coefficient of a single molecule of surfactant. For simplicity, we assume here that the concentration is below the *cmc*, so that the surfactant moves in the form of individual molecules.

How can one evaluate l? By drawing all the surfactant molecules contained within the thickness l below the surface, we ought to generate the surface density Σ^{-1} at equilibrium—a quantity that is relatively well defined for a saturated layer. Hence the condition

$$cl = \Sigma^{-1}. \tag{8.24}$$

Typically, the volume fraction Φ occupied by the surfactant is $\Phi = ca^3 \approx 10^{-3}$, while $\Sigma \approx a^2$, where a is the size of a molecule. From there we conclude that

$$l \approx \frac{a}{\Phi} \approx 0.1\ \mu\text{m}. \tag{8.25}$$

The corresponding time τ [given by equation (8.23)] is of the order of 10^{-4}s.

Conclusion. If the bubble injection takes place in time shorter than τ, the surface tension remains close to that of pure water. These considerations are important in understanding which surfactants can be effective as foaming agents. The criterion is that l should not be too large. In light of equation (8.24) this means that c should not be too small. Stated in words, surfactants that are highly hydrophobic (i.e., with a very low *cmc*) cannot be good foaming agents.

8.3   Some Applications of Surfactants

8.3.1   Flotation

Flotation is a technique that allows us to improve the quality of mineral ores. Raw ores often have two components:

1. Particles containing the metal component (e.g., copper oxide, tin oxide, etc.), which are weakly hydrophilic.
2. A matrix surrounding the particles—often a silicate—which is more hydrophilic.

The standard technique for separating the two is to crush the raw material to create separate grains of types (1) and (2). The mixture is then injected at the bottom of a huge tank filled with water and surfactants. Air bubbles are introduced, and a remarkable phenomenon takes place. The valuable ore is dragged toward the top, while the matrix remains at the bottom!

This process constitutes the most important application (from a tonnage point of view) of surfactants. The details of the operation are rather complex, but the basic ideas can be captured in simple terms.

- *If the bubbles are larger than the grains*, the process can be adjusted so that the metallic particles M attach themselves to the surface of the bubbles (Figure 8.6a).

    This requires that the solution (water + surfactant) be in a partial wetting regime on the surface of M. The corresponding Young's condition in this situation is (see chapter 1)

$$\gamma_{MA} - \gamma_{MW} < \gamma \tag{8.26}$$

where γ_{MA} is the surface energy at the interface between M and air, γ_{MW} is the surface energy at the interface between M and the solution, and γ is the surface tension of the solution itself. If equation (8.26) is satisfied, the particles will gain energy by migrating to the water/air interface.

The anchoring process can fail if the weight $\rho_M g R^3$ of the metal particle (where ρ_M is the density of M, and R is the radius of the particle) is excessive. To prevent this outcome, the capillary forces must prevail over the weight. These forces are of the order of $\bar{\gamma}$ (an average of surface tensions listed above) times the length of the triple line ($\approx R$). Hence, the anchoring condition is

$$\bar{\gamma} R > \rho_M g R^3. \tag{8.27}$$

Introducing once again the capillary length κ^{-1}, the result becomes

$$R < \kappa^{-1} \left(\frac{\rho}{\rho_M} \right)^{1/2} \approx \kappa^{-1}. \tag{8.28}$$

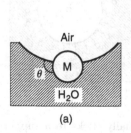 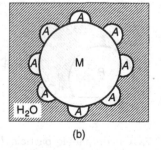

(a)   (b)

FIGURE 8.6. Flotation method for the extraction of an ore.

The conclusion is that only particles smaller than a about millimeter will get anchored. Once stuck to a bubble, such a particle will rise up the height of the tank.

- *Circumstances are also favorable if the bubbles are smaller than the particles.* The condition for partial wetting expressed in equation (8.26) implies that bubbles can attach themselves to M (Figure 8.6b). As long as M does not have too high a weight, it will be dragged up to the top of the tank by a swarm of small bubbles.

What about the matrix particles (G)? The goal here is to have total wetting, which happens when

$$\gamma_{GA} - \gamma_{GW} > \gamma. \tag{8.29}$$

The condition will be met if the matrix is sufficiently hydrophilic. In that case, the matrix particles will fail to anchor themselves to the bubbles and they will remain at the bottom of the tank.

8.3.2 Detergents

The (idealized) problem of detergency is illustrated in Figure 8.7. On a support (S) such as a textile fiber (to take a specific example, let us take polyester), we have a particle P of grease, which we seek to remove. Both substances are hydrophobic. If we attack such a system with a solution of surfactants, its is possible to have the surfactant tails attach themselves to the surfaces of both S and P, which renders them hydrophilic by lowering the interfacial energies $\tilde{\gamma}_{PW}$ and $\tilde{\gamma}_{SW}$ with water by a considerable amount. Suppose we achieve the inequality

$$\tilde{\gamma}_{PW} + \tilde{\gamma}_{SW} < \gamma_{PS} \tag{8.30}$$

where γ_{PS} is the contact energy between P and S (per unit surface). In such a case, we gain energy by separating P from S. In other words, P will then let go of S and will be rinsed away by a hydrodynamic flow.

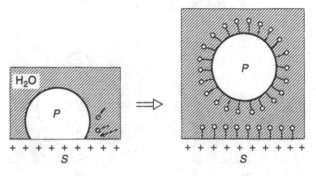

FIGURE 8.7. A hydrophobic particle P initially sticks to a support S. In the presence of a surfactant, it lifts off.

The mechanism we have just described is dominant, but it is not necessarily unique. To it we must also add the following:

- *The possibility of micellization.* A micelle of surfactant inflates itself by capturing in its midst molecules of greasy dirt.
- *The possibility of emulsification.* Instead of an inflated micelle (with a size less than 10 nm), what forms here is a larger drop of grease surrounded by surfactant, with an overall size of about 1 μm).

What type of surfactant is needed for these various functions to be activated?

1. The surfactant must be very soluble in water (therefore, HLB > 7).
2. For the surfactant to break through the S/P interface its aliphatic tails should not be too long. With long chains the penetration process becomes overly sluggish. This demands a high HLB (HLB ≈ 13 – 15).
3. The formation of an emulsion of oil in water requires a high HLB (this will be explained in the next section).

8.3.3 Emulsification

Emulsions are droplets of water in oil (O/W) or of oil in water (W/O), the size of which is typically a few microns. They are produced by forcing mixtures of water, oil, and surfactant through a nozzle, rotor, or other mechanical device. Emulsions have enormous industrial importance. For instance, many active substances are soluble only in non-polar solvents (such as oils) bust must be diluted in water to reduce the toxicity of the pure solvent.

An emulsion is not fully stable. Even in the presence of a surfactant, there is a finite interfacial energy γ_{OW}. The system may lower its energy by recombining the droplets, which decreases the oil/water surface area. Nevertheless, surfactants help since they can lower the energy γ_{OW} by a factor of at least 10. As a rough guideline, a short-lived emulsion requires $\gamma_{OW} = 5$ mJ/m^2. A highly stable emulsion requires a more stringent $\gamma_{OW} < 0.5$ mJ/m^2.

Is oil in water (O/W) more favorable than water in oil (W/O), or the other way around? The answer depends primarily on the particular surfactant used. An empirical rule due to Bancroft states that, if the surfactant is soluble mainly in oil (HLB ~ 3 to 6), it will form W/O emulsions, whereas if it soluble mainly in water (HLB ~ 8 to 18), it will form O/W emulsions.

The physical origin of Bancroft's rule is rather uncertain. Three explanations will be discussed here.

- *On a scale of 10 nm* (Figure 8.8), surfactants with a small head (i.e., with a low HLB) arrange themselves more readily on a W/O interface than on a O/W one. However, this argument is debatable. Emulsion drops are large, actual curvatures are quite weak, and the energy difference

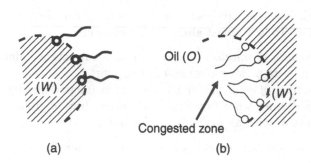

FIGURE 8.8. Argument used to justify Bancroft's rule.

between the W/O and O/W configurations resulting from this effect is negligible. This explanation is unsatisfactory.

- *Low nucleation energy of bridges between droplets.* Curvatures can play a more important role during a coalescence process, i.e., when two oil droplets in water draw near each other. In this scenario, the droplets can coalesce, establishing a bridge of surfactant-coated oil. Such a coating is under favorable conditions if the aliphatic tails are larger than the heads. Coalescence is easier at low HLB for an O/W emulsion, whereas it is difficult at large HLB. As a result, O/W emulsions are more stable in the latter case, in agreement with Bancroft's rule (we owe this argument to D. Roux).

- *On the scale of a drop,* interesting effects related to the transport of surfactant toward the interface can occur (Figure 8.9). Consider, for example, the formation of a finger of water in the oil region as a result of agitation. If the surfactant is soluble in oil, it quickly attaches itself to the newly formed interface and lowers its surface tension. If so, the finger can persist for a long time (ultimately, it will break up into W/O droplets due to the Rayleigh instability). By contrast, if the surfactant is soluble in water, it would have a difficult time making its way into the finger (the distance to cover is large). The surface tension would then remain high and the finger would retract swiftly.

8.3.4 Surfactants as Wetting and Dewetting Agents

It is often desirable to spread a thin film of solution (W) on a hydrophobic surface (S). This does not happen spontaneously. If one tries to spread a thin film in a partial wetting regime, the film tends to fragment in droplets, each of which conforms to the contact angle θ_E specified by Young's relation. As discussed in chapter 7, only thicker films, with a puddle thickness $e_c = 2\kappa^{-1}\sin(\theta_E/2)$ in the millimeter range, are stable. One must therefore resort to an additive to lower θ_E (and hence e_c). In practice, this additive is a surfactant. The products we use to clean our kitchens are formulated

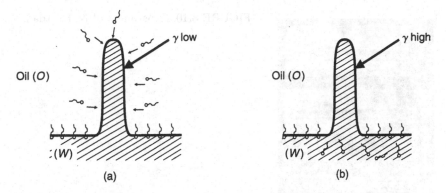

FIGURE 8.9. Third argument used to justify Bancroft's rule: Formation of a finger destabilizing into droplets.

in such a way as to decrease e_c, making housecleaning chores much less strenuous.

Let us reexamine the condition for total wetting derived in chapter 1. It reads

$$\gamma + \gamma_{SW} < \gamma_{SA}. \tag{8.31}$$

To meet this criterion, we can decrease γ by means of a surfactant soluble in water. Two cases must be distinguished.

1. If the surfactant is anionic (negative ion) and if S is a mineral surface charged negatively, the two will repel each other and the surfactant cannot attach itself to S. As a result, the energy γ_{SW} remains unchanged and the surfactant acts on the surface tension γ only.
2. With other solids, additional tensoactive effects can exist at the S/W interface. For instance, if the surface is non-polar, the surfactant can affix itself onto it by its aliphatic part, lowering the energy γ_{SW}. This case is reminiscent of the detergent problem. One could argue that the surfactant eliminates air bubbles much as a detergent eliminates grease particles.

What kind of surfactant makes a good wetting agent? It must be highly soluble in water (the *cmc* should not be too low, meaning HLB > 7). On the other hand, it need not be able to squeeze between two solid surfaces. Therefore, the requirement that the aliphatic tail be short, which we had identified in the case of detergents, no longer applies. In practice, we need a surfactant with an HLB in the neighborhood of 7.

As shown in section 8.3.2 on detergents, surfactants can also promote the *dewetting* of a liquid on a solid. That is indeed the very principle behind a detergent. Figure 8.10 reproduces a sequence of photographs taken by N. K. Adam—a pioneer in the physical chemistry of surfaces, who worked in the 1920–1950 period. The figure illustrates how a film of oil deposited on a thread evolves in time after being immersed in soap water. The oil,

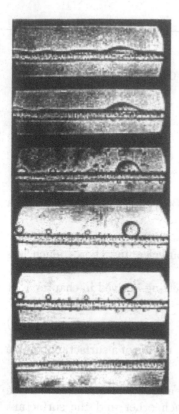

FIGURE 8.10. Experiment of N. K. Adam.

wetting at first, becomes increasingly dewetting until it reaches a stage characterized by a contact angle of 180° (zero wetting), at which point adhesion to the substrate ceases. Droplets come loose, and the thread ends up completely clean.

8.4   Soap Films and Bubbles

8.4.1   Fabrication of Films

The study of soap films goes back to Plateau (1873). It reached its peak with Karol Mysels (1959).[3]

A simple way to make soap films is to dip a rigid frame in a solution of soap water and surfactant and to pull it out as shown in Figure 8.11.

Success hinges on several precautions:

1. One must avoid the evaporation of water and work in a closed container.

2. It is sometimes advantageous to inhibit water from draining away toward the bottom by increasing the viscosity with added glycerin.

FIGURE 8.11. Drawing a soap film
with a frame.

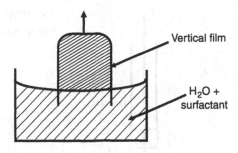

Films obtained with this method typically have a thickness of the order of a micron. They produce selective reflection at particular optical
wavelengths. When observed in white light, they display colors which are
characteristic of their thickness e.

The films are thinner at the top than at the bottom. After a while,
the upper part often ends up becoming extremely thin, reducing to two
soap monolayers sandwiching just a few molecules of water. Such films no
longer produce optical interference effects; instead, they look quite black.
Newton studied these black films. Long before, the Assyrians had observed
them and even used their unpredictable shapes (in horizontal geometries)
to divine the future.

Throughout the remainder of this chapter, we will restrict ourselves to
"thick" films (about a micron). In this case, long-range forces can be ignored
and macroscopic concepts such as the surface tension $\gamma = \gamma_0 - \Pi$ do apply.

8.4.2 The Role of Surfactants

How does the film manage to remain suspended in air despite its weight?
This is explained in Figure 8.12.

Monolayers of surfactant exist on the either side of the film with a certain
pressure $\Pi(z)$ at altitude z. A horizontal slice from z to $z + dz$ has a weight
$\rho g e\, dz$, where e is the thickness of the film. This weight is balanced out by
a gradient in the surface pressure, created by a denser surfactant (higher
Π) at the bottom than near the top. This translates into the condition

$$\rho g e = -2\frac{d\Pi}{dz} \tag{8.32}$$

where the factor of 2 accounts for the two monolayers—one on either side.

It now becomes clear why a surfactant is necessary. We can also predict
over what thickness range the film will be able to hold up on its own.
Let us assume that the concentration of the solution is above the *cmc*. In
accordance with equation (8.16), the corresponding pressure Π is equal to
the surface tension γ_0 of pure water, minus the surface tension γ_∞ of the
saturated solution. For a rough estimate, we may assume that the thickness

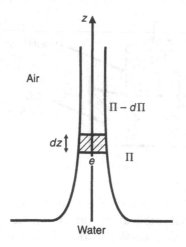

FIGURE 8.12. Equilibrium of a "young" film of soap water.

Air

$\Pi - d\Pi$

$dz\updownarrow$

e

Π

Water

e is constant. The maximum height $z_{\max}$ of the film is then obtained by integrating equation (8.32). The result is

$$\rho g z_{\max} e = 2\left(\gamma_0 - \gamma_\infty\right).\qquad(8.33)$$

For $e = 1$ μm and $\gamma - \gamma_\infty = 50$ mJ/m^2, one finds that $z_{\max}$ is of the order of 10 m. In reality, the film is thinner at high altitudes, and $z_{\max}$ is even larger. A more complete analysis can be found elsewhere.[4]

8.4.3   Draining Mechanisms

A film slims down by expelling its water content toward the bottom. This process is referred to as "draining."

- *Condition (8.32) is necessary but not sufficient.* The weight is applied by the water, and the resistance is exerted by the surfactant. But water and surfactant are not rigidly tied to each other. Water can drain between the two fixed walls via a Poiseuille flow across the thickness e. However, the corresponding velocities are quite slow because the thickness e is small. In practice, we can neglect this direct draining mechanism.
- *There exist other mechanisms for liquid transfer.* The basic idea is illustrated in Figure 8.13, where one region of the film is thinner than the rest. This region is consequently also lighter. The situation is quite analogous to that of a hot air balloon, which contains a light gas (hot air) and is immersed in a heavier gas (cold air). The thin region of the film will rise much as does the balloon propelled by Archimedes' pressure.
- *When a thin film is attached to a vertical frame,* the regions immediately adjacent to the frame are constricted. The reason is illustrated in Figure 8.14. A molecule at point A is at atmospheric pressure. One at point B experiences a Laplace pressure, which is negative (lower than the atmosphere's). Under these circumstances, a flow is set up from A to B, which

FIGURE 8.13. Archimedes' pressure exerted on a thinner region of the film.

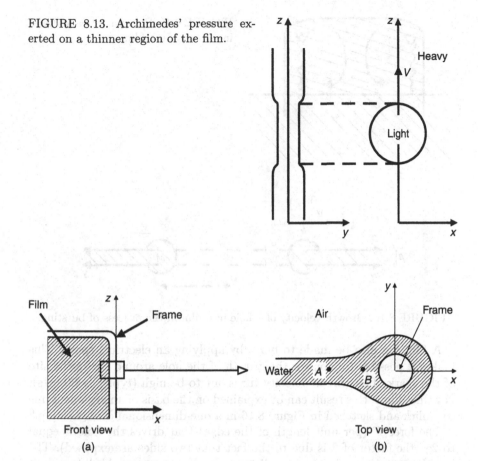

FIGURE 8.14. Thinning down of a film near the frame.

results in a thinning of region A.[5] A strip of film near the frame ends up being lighter and tends to rise.

This is akin to a vertical radiator heating a room. Warm air near the radiator starts rising and generates turbulent flows. Similar turbulent flows exist in a soap film near the lateral walls. These flows close the loop at the center (Figure 8.15). Thus, there is a net redistribution of liquid from the central portion of the film toward the bottom. The physical laws involved are rather complex. For more details, the reader is encouraged to consult the book by Mysels, Shinoda, and Frankel.[3]

8.4.4 Aging and Death of Films

In the mechanism we have just described, the films thin down in their upper portion—the region of the so-called "black film." They eventually burst, typically by way of a hole nucleating around a speck of dust.

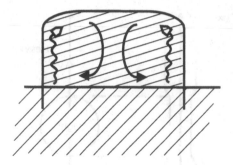

FIGURE 8.15. Flow pattern and draining process in a film.

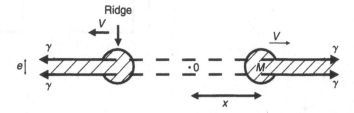

FIGURE 8.16. Growth velocity of a hole in a film in the process of bursting.

A film can also be made to burst by applying an electrical spark. This makes it possible to monitor the growth of the hole around the initial site of the spark. The growth velocity turns out to be high (typically 30 m/s). For thick films, this result can be explained on the basis of an argument due to Culick and sketched in Figure 8.16 in a one-dimensional configuration.[6]

The force F (per unit length of the edge) that drives the hole is equal to 2γ (the factor of 2 is due to the fact that two sides are exposed). The liquid is set in motion over a distance x. The mass involved (per unit length) is $M = \rho e x$. If this entire mass is transported at a velocity V, the corresponding momentum is

$$P = MV. \tag{8.34}$$

Newton tells us that $F = \frac{dP}{dt}$. In usual exercises from high school physics, M is constant and V is variable. Here, instead, the situation is reversed: V turns out to be constant and M is variable. We have:

$$\frac{dM}{dt} = \rho e \frac{dx}{dt} = \rho e V. \tag{8.35}$$

From this we derive

$$\rho e V^2 = 2\gamma. \tag{8.36}$$

Solving for V gives

$$V = \left(\frac{2\gamma}{\rho e}\right)^{1/2}. \tag{8.37}$$

This velocity is high (10 m/s for $e = 1\,\mu\mathrm{m}$). The phenomenon is much too rapid to be followed by the naked eye, but it can be captured by high-speed photography.

Culick's argument is rather subtle for several reasons:

1. We do not know precisely what shape the water ridge takes, nor do we know whether it breaks up into droplets because of the Rayleigh instability. What becomes of the excess surfactant is not known either.

2. Energy is not conserved in the proposed mechanism. Therefore, there must be some form of dissipation. If we were to write that all of the capillary energy $2\gamma S$ (where S is the surface area of the film) is converted into kinetic energy $1/2MV^2$, we would still obtain a constant bursting velocity, but with a numerical coefficient different from the one appearing in equation (8.37).

3. Given their high pressure Π, monolayers are rather rigid. As such, they can transmit acoustic signals (shock waves) at velocities superior to V. This translates into the appearance of *aureoles* around the hole, as discussed by Mysels and Frankel.[7] On the whole, the dynamics of rupture is relatively clear for thick films (one micron or more). It is much more complex for thin films, where surfactant effects become predominant.

8.4.5   The Case of Bubbles

To a first approximation, a bubble has a spherical shape with a radius R and a constant thickness e. Upon further analysis, however, it becomes clear that the equilibrium of a bubble is a bit more subtle. Our discussion will assume, here again, that direct draining is negligible (Figure 8.17).

• Just as is the case in ordinary films, there are gradients $\nabla\Pi$ of surface pressure to balance out the weight of the liquid. We can write down an equilibrium condition involving tangential forces. The gradient of the Laplace pressure $(2\gamma/R)$ must be equal to the tangential component of the weight:

$$\frac{2d\gamma}{Rd\theta} = -\rho g \sin\theta \tag{8.38}$$

where θ is a polar angle. From this we derive

$$\gamma(\theta) = \gamma|_{\theta=\pi/2}\left[1 + \frac{1}{2}\kappa^2 eR\cos\theta\right]. \tag{8.39}$$

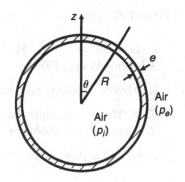

FIGURE 8.17. Detailed structure of a bubble: The thickness e depends on θ.

With $R = 1$ cm, $\kappa^{-1} = 2$ mm, and $e = 1$ μm, the correction term is small, of the order of 10^{-3}.

- If we write down the equilibrium of forces in a direction normal to the surface of a bubble, we get:

$$p_i - \frac{2\gamma}{R} + \rho g e \cos\theta = p_e + \frac{2\gamma}{R} \qquad (8.40)$$

where p_i is the internal pressure within the bubble, and p_e is the external (atmospheric) pressure. This leads to

$$\frac{4\gamma}{R} - \rho g e \cos\theta = p_i - p_e = cnst. \qquad (8.41)$$

The variation of γ with latitude [equation (8.39)] must be inserted into equation (8.41) to give finally

$$\frac{4\gamma}{R(\theta)} + \rho g e \cos\theta = p_i - p_e. \qquad (8.42)$$

Conclusions.

- The internal pressure is higher than p_e, in agreement with Laplace's law. The $\rho g e$ term in equation (8.42) constitutes a small correction.
- The radius of curvature depends slightly on the latitude (through the angle θ). The curvature is minimum at the top of the bubble. However, this correction is small, (of the order of 10^{-2} or less) since we have

$$\frac{\Delta R}{R} \approx \kappa^2 e R.$$

In practice, a bubble is not quite the ideal object we have just described. At the moment of its birth, the thickness e is not the same everywhere. There are heavier regions and lighter regions. This gradient can generate flows akin to turbulent motions in films and gives rise to the shimmering iridescence of bubbles.

References

[1] L. D. Landau and E. M. Lifshitz, *Statistical Physics* (Reading, Mass.: Addison-Wesley, 1958).

[2] K. Mysels, *Colloids and Surfaces* **43**, 241 (1990).

[3] K. J. Mysels, K. Shinoda, and S. Frankel, *Soap Films: Studies of their Thinning, and a Bibliography* (New York: Pergamon Press, 1959).

[4] P. G. de Gennes, *Langmuir* **17**, 2716 (2001).

[5] A. Aradian, E. Raphaël, and P. G. de Gennes, *Europhys. Lett.* **55**, 834 (2001).

[6] F. E. Culick, *J. Appl. Phys.* **31**, 1128 (1960).

[7] S. Frankel and K. Mysels, *J. Phys. Chem.* **73**, 3028 (1969).

9
Special Interfaces

9.1 Outline

This chapter addresses three topics, each with great practical importance.

- The *wetting of textured surfaces* describes not only the wetting of duck feathers but also that of some complex structures the industry is beginning to produce. We have seen in chapter 2 that a standard method of altering the wettability of a surface is to treat it chemically by grafting or adsorbing molecules with wetting characteristics of their own. Another method routinely used by nature is to texture the surface, in other words, to modify the surface roughness. Such textures can be either mildly disordered or strongly disordered and even (to some extent) fractal. If the spatial scale of the roughness is small compared to the wavelength of light the texture does not affect optical transparency. This opens the door to the development of "super hydrophobic" windshields, which would combine standard chemical treatments with a surface structure on an ultra-fine scale. We will describe the static and dynamical properties of drops that hop, glide, and roll on such surfaces.

- The *wetting of porous media* describes the properties of another category of very rough surfaces, albeit with a new twist. The three-dimensional nature of the roughness gives rise to the phenomenon of impregnation. A drop placed on a porous medium does not merely spread on the surface but also penetrates the depth of the support, thereby modifying its wetting properties. A liquid film deposited on a porous substrate recedes via two separate mechanisms: suction and dewetting. This topic encompasses the wetting of powders, rocks and soils, and the paper industry.

- In a third part dealing with the *wetting of soft interfaces*, air is replaced by a deformable medium, namely, a rubber. We shall see that the static and dynamical wetting properties are then completely different because the deformation energy of the elastomer has to be taken into account. The competition between surface and elastic energies introduces a new characteristic length h_0, called the *elastic length*, (which ranges typically in the tens of nanometers). Ordinary capillarity applies only if the spatial scales involved are less than h_0. We shall examine the shape of drops and discuss how sandwiched films recede. This phenomenon governs the adhesion of objects on wet substrates (automobile tires, boat glues, contact lenses), as well as certain processes used in the printing industry. In biology, it may play a role in the formation of adhesive contacts between living cells.

9.2   Wetting of Textured Surfaces

9.2.1   Basic Model

9.2.1.1   Experiment of Johnson and Dettre

In practically every case we have discussed up to now, the solid surface was assumed to be ideal, that is to say, smooth and chemically homogeneous. The contact angle θ_E of a drop on such a solid is given by the now familiar Young relation, which is reproduced here:

$$\cos\theta_E = \frac{\gamma_{SV} - \gamma_{SL}}{\gamma_{LV}} \tag{9.1}$$

where γ_{IJ} designates the surface tension (or energy) between phases I and J.

In practice, of course, solids are typically both heterogeneous and rough. The two often reinforce each other. Fissures at the surface can expose a different medium or promote corrosion, which is apt to alter the chemical composition of the surface at that particular spot. It is therefore important to gain some understanding of how these defects can influence wetting. We have already underscored that one of the consequences of these defects is a hysteresis of the contact angle, covering a range of values that can be as large as several tens of degrees around a mean.

Conversely, we know that it is possible to alter the wetting properties of a substrate by changing the texture of its surface. In this context, it is worth reviewing the historic experiment of Johnson and Dettre in 1964, who measured the contact angles (both advancing and receding) of water drops placed on wax surfaces.[1] Successive heat treatments enabled them to vary the degree of roughness of the substrate, (which became smoother after each bake), while maintaining the chemical homogeneity of the surface.

The results obtained by Johnson and Dettre are reproduced in Figure 9.1. Ideally the contact angles, expressed in degrees, should be plotted as

FIGURE 9.1. Static contact angles (advancing and receding) of water drops placed on wax surfaces as functions of the roughness of the substrate. The horizontal scale is qualitative; changes in roughness are achieved by successive bakes (reproduced from ref. 1).

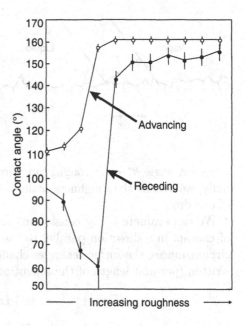

functions of the substrate roughness r *defined as the ratio of the real surface area to the apparent surface area.* Johnson and Dettre did not rigorously measure this parameter. As a result, the abscissa is qualitative only.

Evidently, the surface roughness has a considerable influence on both the contact angle itself and its hysteresis. At low roughness, the advancing angle increases with r, whereas the receding angle decreases (therefore, the hysteresis increases in this first regime). Beyond a certain value of r, both angles increase rather sharply. At the same time, the hysteresis drops to a very small value. From a practical standpoint, texturing would then appear to be a particularly effective way to tailor the wetting characteristics of a surface.

We are interested here mostly in the average value of the contact angle, and we will examine how it deviates from Young's value for a flat surface as a function of the degree of texturing. We restrict our attention to those cases where the hysteresis of the contact angle is small. We begin by presenting the standard wetting models of surfaces that are textured either physically (Wenzel's model) or chemically (Cassie-Baxter model). Next, we will show how these two models can be combined in many hybrid cases (composite surfaces). Finally, we shall describe a particularly remarkable case of wetting of a textured surface, namely, *zero wetting.*

9.2.1.2   Wenzel's Model

One the first attempts at understanding the influence of roughness on wetting is due to Wenzel (1936).[2] We assume that the *local* contact angle is given by Young's relation [equation (9.1)], and we seek to determine the

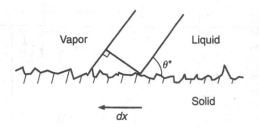

FIGURE 9.2. Edge of a drop placed on a rough surface. We consider a small displacement dx (toward the left) of the line of contact.

apparent angle θ^* on a rough, but chemically homogeneous, surface. Finally, we assume the roughness scale to be very much smaller than the size of our drop.

We can evaluate θ^* by considering a small displacement dx of the line of contact in a direction parallel to the surface (Figure 9.2). Under these circumstances, the surface energies change by an amount dE, which can be written (per unit length of the line of contact) as

$$dE = r(\gamma_{SL} - \gamma_{SV})\, dx + \gamma_{LV}\, dx \cos\theta^* \qquad (9.2)$$

where r is roughness of the solid. The minimum of E defines equilibrium. For $r = 1$ (smooth surface), we do indeed recover Young's relation (9.1). By contrast, when $r > 1$, the equilibrium condition leads to Wenzel's relation:

$$\boxed{\cos\theta^* = r\cos\theta_E} \qquad (9.3)$$

where θ_E is Young's angle.

Wenzel's relation embodies two types of behavior:

1. If $\theta_E < 90°$ (hydrophilic solid), we will have $\theta^* < \theta_E$ since $r > 1$.
2. Likewise, if $\theta_E > 90°$, we will have $\theta^* > \theta_E$.

Surface roughness always *magnifies the underlying wetting properties*. Both hydrophilic and hydrophobic characteristics are reinforced by surface textures. Equation (9.3) also predicts wetting and drying transitions. Since the roughness r is not bounded, there should exist a threshold value r^* beyond which wetting becomes either total or zero, depending on the sign of $\cos\theta_E$. This threshold value, given simply by $r^* = 1/\cos\theta_E$, is a priori easily accessible. For $\theta_E = 60°$, we have $r^* = 2$. We will, however, see later on that this statement is highly arguable. We shall specify (and restrict) the domain of validity of Wenzel's relation.

9.2.1.3   The Cassie-Baxter Model

The same line of reasoning can be applied to a surface that is *planar but chemically heterogeneous*.[3] We will now assume the surface to be made of two species (Figure 9.3), each characterized by its own contact angle θ_1 and

FIGURE 9.3. Edge of a drop placed on a chemically composite surface.

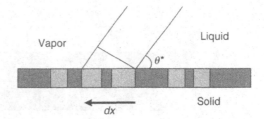

θ_2, respectively. We will denote by f_1 and f_2 the fractional surface areas occupied by each of these species ($f_1 + f_2 = 1$), and we will assume again that the individual areas are very small compared to the size of a drop.

Letting θ^* denote, here again, the apparent angle, the energy variation associated with a small displacement dx is

$$dE = f_1(\gamma_{SL} - \gamma_{SV})_1\, dx + f_2(\gamma_{SL} - \gamma_{SV})_2\, dx + \gamma_{LV}\, dx \cos\theta^* \qquad (9.4)$$

where the indices 1 and 2 refer to one or the other species swept during the displacement—species 1 with a probability f_1, and species 2 with a probability f_2. The minimum of E, together with Young's relation applied to each solid, lead to the Cassie-Baxter relation:

$$\cos\theta^* = f_1\cos\theta_1 + f_2\cos\theta_2 \qquad (9.5)$$

Therefore, the apparent angle (which is indeed restricted to the interval $[\theta_1, \theta_2]$) is given by an average involving the angles characteristic of each constituent, but the average is applied to the cosines of these angles. For a review of the behavior of drops on chemically textured surfaces, the reader should consult a recent paper.[4]

9.2.2   Composite Rough Surfaces

We now return to the topic of rough surfaces and examine Wenzel's approach in a more critical light. We shall consider in succession the case of a hydrophilic surface ($\theta_E < 90°$) and that of a hydrophobic surface ($\theta_E > 90°$). All this is based on the model of Bico et al.[5]

9.2.2.1   Hydrophilic Surfaces

A very rough hydrophilic surface is somewhat similar to a thin porous medium. We can imagine that some liquid escapes from the drop and penetrates into the nooks and crannies of the solid. The volume of film captured in the recesses is generally negligible, and in the end the drop finds itself essentially on a wet substrate viewed as a patchwork of solid and liquid.

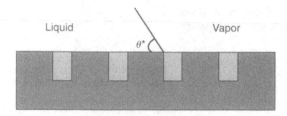

FIGURE 9.4. Hydrophilic porous surface. If the liquid spontaneously fills the grooves, a drop will find itself on a mixed solid/liquid surface. Equilibrium imposes a nearly vanishing Laplace pressure, implying a horizontal surface.

In the interest of simplicity, we restrict our attention to a surface made of an array of well-defined tiles or grooves. Suppose that, as depicted in Figure 9.4, the grooves are filled with a liquid and the plateaus are dry (we are in a partial wetting regime). Our objective is to calculate the apparent angle θ^* of a drop deposited on such a surface.[5]

We denote by Φ_S and $1 - \Phi_S$ the relative fractions of the solid and liquid phases underneath the drop. The Cassie-Baxter relation [equation (9.5)] can be applied to the mixed surface, with contact angles θ_E and 0, respectively. We then deduce θ^*:

$$\cos\theta^* = 1 - \Phi_S + \Phi_S \cos\theta_E. \tag{9.6}$$

It becomes clear that, contrary to the predictions of Wenzel's law, we cannot induce total wetting ($\theta^* = 0$) simply by means of a surface texture. Since we are in a partial wetting regime ($\theta_E \neq 0$), there will always remain islands that emerge above the "absorbed" film.

The condition for this description to be valid is for a penetrating film to develop. As it turns out, the criterion for the existence of such a film is more restrictive on a rough surface than on a standard porous material. Indeed, on a rough surface, the liquid film involves a free interface, which costs energy. We would then expect the criterion for penetration in two dimensions to be intermediate between total wetting ($S > 0$ and $\theta_E = 0°$) and standard absorption ($\theta_E < 90°$).

We may think of the surface as an array of tiles and consider that the wetting film advances by a distance dx on such a surface. During such an event, the wet surface area (per unit length of the film) is $r\,dx$, but the movement also leaves a solid surface area $\Phi_S\,dx$ dry. The corresponding energy variation (still per unit length) is then given by

$$dE = (r - \Phi_S)(\gamma_{SL} - \gamma_{SV})\,dx + (1 - \Phi_S)\gamma_{LV}\,dx. \tag{9.7}$$

We see that the movement of the line is energetically favorable ($dE < 0$) if the cosine of Young's angle verifies the inequality

$$\cos\theta_E > \frac{1 - \Phi_S}{r - \Phi S}. \tag{9.8}$$

Equation (9.8) defines an angle θ_c such that, when $\theta_E < \theta_c$, a film will impregnate the texture. As expected, θ_c is always between 0 and 90°.

We can also derive the value of θ_c by calculating the intersection of the curves described by equations (9.3) and (9.6). This helps us to understand that the two laws must be obeyed in turn as $\cos\theta_E$ increases. When Young's angle is between θ_c and 90°, the solid remains dry ahead of the drop, and Wenzel's law describes wetting correctly. When it is between 0° and θ_c, a film penetrates the texture and the drop rests on a solid/liquid composite, in which case equation (9.6) is the one that describes wetting correctly.

9.2.2.2 Hydrophobic Surfaces

Calculation of the Contact Angle

We now reexamine Wenzel's relation in the case of a hydrophobic rough surface. To say that a material is hydrophobic means that the surface energy of the dry solid is less than that of the wet solid, consistently with Young's relation (in which $\theta_E > 90°$ is equivalent to $\gamma_{SV} < \gamma_{SL}$). Under these conditions, we would not necessarily expect the solid/liquid interface to conform to the topographical features of the solid surface, contrary to what was implicitly assumed in deriving Wenzel's relation. *Air can actually remain trapped under the drop*, at least as long as Young's relation remains satisfied wherever contact lines appear. If so, the drop rests on a composite surface made of both solid and air. Our objective is to establish the conditions for such air pockets to appear and to determine the resulting macroscopic contact angle.

We begin by considering the value of the apparent contact angle θ^* adopted by the system. Once again, we consider the simplest possible arrangement of solid and air, which consists of an array of holes in a solid surface, where the various interfaces are all planar, as shown in Figure 9.5. As before, Φ_S denotes the fraction of solid underneath the drop.

Again, the surface is mixed and we can resort to the Cassie-Baxter relation [equation (9.5)] to determine the angle θ^*. The two phases involved (solid and air) are characterized by their respective contact angles θ_E and π,

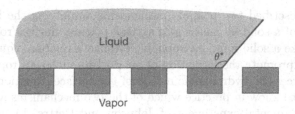

FIGURE 9.5. Hydrophobic porous surface. The liquid does not necessarily fill the pores, and a drop rests on a composite of solid and air.

and occupy fractional surface areas Φ_S and $1 - \Phi_S$. We deduce the relation

$$\cos \theta^* = -1 + \Phi_S(\cos \theta_E + 1). \qquad (9.9)$$

Equation (9.9) implies a very different behavior from that predicted by Wenzel. It reveals that as soon as air pockets are present, the cosine of the contact angle jumps to a value less than or equal to $\Phi_S - 1$, ultimately approaching the value -1 (contact angle of 180°) provided that the fractional area of the solid is small. It also suggests that the ideal value of 180° can never be reached, since the only ways to achieve this limit are $\Phi_S = 0$ or $\theta_E = 180°$, neither of which is physically realizable. This is in contrast to Wenzel's law, which predicts that the angle is pinned at 180° the moment the product $r \cos \theta_E$ reaches the value -1. It is difficult in practice to prepare surfaces for which Young's angle exceeds 120° ($\cos \theta_E = -0.5$). Expanding equation (9.9) when $\theta^* \to \pi$ shows that the apparent angle will be hard pressed to go beyond a value of the order of $\pi - \sqrt{\Phi_S}$ (equal to 162° when $\Phi_S = 10\%$). This function is critical (in the sense that it varies very rapidly) in the vicinity of $\Phi_S = 0$. This highlights the difficulty in realizing super-hydrophobic surfaces.

We should note that the simplicity of our reasoning stems from having assumed all interfaces in Figure 9.5 *coplanar*. The analysis can be complicated somewhat by retaining some degree of roughness of the solid part. The liquid/vapor interfaces, on the other hand, can never deviate very much from planarity. Indeed, the size of a drop is typically a millimeter, whereas surface defects are on a scale of a micron. Since the Laplace pressure must be constant at equilibrium, it is legitimate to consider these interfaces planar.

We must finally discuss the stability of the air-trapping regime.[6] Is it energetically more favorable to conform to the surface roughness (Wenzel's model) or to trap air inside the solid texture? If we compare the energy variation associated with a small displacement of the contact line in both cases, we find that it is more favorable to follow the roughness if the contact angle θ_E is between 90° and a certain angle θ_c given by $\cos \theta_c = (\Phi_s - 1)/(r - \Phi_s)$. Experiments show that even in this interval, air is still often trapped (see, for instance, Figure 9.10). Thus, this regime is metastable. For angles θ_E larger than θ_c, the regime is stable.

Promotion of Trapping

We have described two possible mechanisms amplifying the hydrophobic properties of a solid by means of a surface texture. Surface roughness can readily make a solid more hydrophobic [Wenzel's relation, equation (9.3)]; it can also promote the formation of air pockets under a drop, which further reinforces the hydrophobic nature of the surface [equation (9.9)]. It is important to know in practice which of the two mechanisms applies.

The fundamental experiment of Johnson and Dettre (Figure 9.1) indicates that a single scenario is not necessarily sufficient to explain how roughness influences the wetting properties of a solid. The data suggests two distinct behaviors depending on the value r of the roughness. For a small roughness, the advancing angle increases monotonically with r, whereas the receding angle decreases. This would appear to be consistent with Wenzel's law if one considers that the receding angle at zero roughness is less than 90°, so that roughness can only cause it to decrease. At a certain threshold roughness, both angles suddenly jump to a very high value. This transition could be interpreted as the sudden formation of air pockets at that particular roughness (and beyond). The hysteresis of the contact angle, which reflects the degree of heterogeneity of the solid, then decreases by a large amount, even though the roughness itself increases! This paradox is easily resolved if one considers that the substrate actually becomes homogenized as soon as it is composed primarily of air.

We can generalize this argument. Very rough surfaces are more likely to obey equation (9.9), while less rough surfaces will continue to conform to Wenzel's law (in the hydrophobic case). Which law will prevail is determined by the possibility of locally satisfying Young's relation—a necessary condition for air pockets to form. Figure 9.6 compares two sinusoidal surfaces of equal wavelengths but of different amplitudes. Young's angle is chosen to be slightly more than 90°. It is easy to appreciate that the formation of air pockets is possible on the rougher surface only.

If $z = a\cos(kx)$ describes the profile of the solid surface, we can work out a one-to-one correspondence between the value of the roughness and the threshold for air pockets formation. The maximum slope of the profile is ak (in absolute value), and the condition for air trapping (that is to say, the establishment of a horizontal liquid/vapor line of contact) for a given θ_E reads

$$a > \frac{\lambda \tan \theta_E}{2\pi} \tag{9.10}$$

where we have introduced the wavelength λ of the surface ($k = 2\pi/\lambda$). For $\theta = 120°$ and $\lambda = 4\ \mu m$, trapping will be possible as soon as we deal with relief amplitudes of the order of a micron. Furthermore, in the limit of $|ka| \ll 1$, the roughness is simply given by the relation

$$r = 1 + \frac{(ka)^2}{4}. \tag{9.11}$$

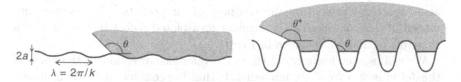

FIGURE 9.6. Ideal solid surface with a sinusoidal profile. For low amplitudes, the liquid is able to follow the waviness. Beyond a certain amplitude threshold, air pockets can form underneath the liquid.

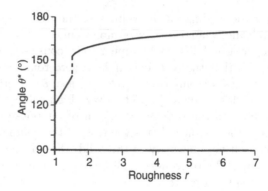

FIGURE 9.7. Apparent contact angle on an ideal rough surface as a function of roughness. There is a discontinuity of the contact angle at the threshold for air pocket trapping.

Combining equations (9.10) and (9.11) yields a threshold roughness r^* for air trapping:

$$r^* = 1 + \frac{\tan^2 \theta_E}{4}. \tag{9.12}$$

For $\theta_E = 120°$, the formation of air pockets will occur for $r > r^* = 1.75$. More generally, we can also calculate for such surfaces the roughness r for all values of ka. For low roughness, the solid/liquid interface conforms to the profile of the solid surface, and the contact angle is given by Wenzel's law. Beyond a threshold r^*, air pockets are trapped and equation (9.9) must be used to evaluate θ^*. We can calculate for each value of r the fraction of the solid in contact with the liquid. Figure 9.7 shows the result of such a calculation for $\theta_E = 120°$.

The shape of the curve $\theta^*(r)$ is qualitatively similar to the data of Johnson and Dettre for the advancing angle (see Figure 9.1). At first, the advancing angle increases monotonically with roughness. It then suddenly jumps at the trapping threshold. Beyond that point, the contact angle continues to increase with roughness, although much more slowly.

These comments apply to the case of solids with a so-called "soft" roughness, where the shape of the solid surface can be described in terms of a function that is continuously differentiable, such as a cosine. In other cases, the solid features discontinuous changes in slope. Profiles of this type are also capable of promoting the trapping of air pockets.[7] To become convinced of this, one need only consider an isolated defect in the form of a slope discontinuity, as shown in Figure 9.8.

We will use the horizontal axis as the reference direction. If the angle of the defect is Ψ, we can see immediately that the contact angle at the defect can take values between Young's angle θ_E and $\pi - \Psi + \theta_E$. If we consider a crenellated profile or a surface with abrupt depressions (as in Figure 9.5), where the angle Ψ is equal to $\pi/2$, the contact angle can deviate from a

FIGURE 9.8. Trapping of the line of contact at the apex of a physical defect. At that location, the contact angle can have any value between θ_E and $\pi - \Psi + \theta_E$.

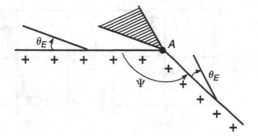

right angle by an amount between θ_E and $\pi/2 + \theta_E$. If θ_E itself is greater than $\pi/2$, an angle π is one possible solution (see Figure 9.5). We can now readily understand why geometric textures with jagged profiles and slope discontinuities have the ability to trap air, and therefore to magnify the hydrophobic properties of a substrate.

9.2.2.3 Summary

We have summarized in Figure 9.9 all our results relative to a surface whose texture promotes the trapping of air. We further assume that the surface is sufficiently simple for the previous discussion to be valid (crenellated profile). In this case, equation (9.9) describes the hydrophobic regime, whereas the hydrophilic regime obeys equations (9.3) and (9.6), depending on the value of the contact angle.

For less ideal textures, the above discussion remains qualitatively correct, as attested by the experimental results obtained by Shibuichi et al.[8] The results are reproduced in Figure 9.10, which uses the same coordinate system as Figure 9.9, for three different surface roughnesses.

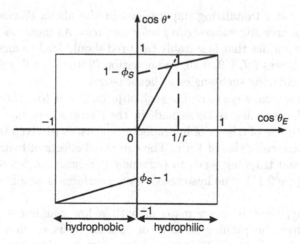

FIGURE 9.9. Apparent contact angle as a function of Young's angle (via their cosines) for a liquid on a textured surface.

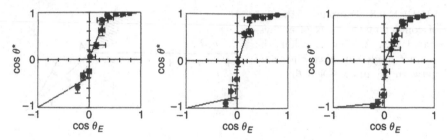

FIGURE 9.10. Experimental data showing the contact angle as a function of Young's angle (measured on a planar surface) for different substrates with random roughness (increasing from left to right).

Although the data were obtained with very complex surfaces (they consist of disordered arrays of micron-size spheres), they match the main features displayed in Figure 9.9. We should emphasize the asymmetry between the hydrophilic ($\cos\theta_E > 0$) and hydrophobic ($\cos\theta_E < 0$) sides of the plots. In the first case, we can see that Wenzel's law ceases to apply when the contact angle becomes too small (when an impregnation film develops), and that indeed we cannot induce total wetting with textures (even in the limit of ultra-rough surfaces of the type used in these experiments). In the second case, we observe a discontinuous jump of the contact angle as soon as Young's angle exceeds 90°. The greater the roughness, the more pronounced the jump. This is to be expected if one considers that a high roughness results in a small solid factional area Φ_S.

9.2.3 Liquid Pearls and Marbles

9.2.3.1 Implementation

One of the most tantalizing implications of the above discussion is the possibility of creating *super-hydrophobic surfaces*. As mentioned before, a hydrophobic surface that is suitably textured should lead to angles close to 180° ([see Figures 9.7, 9.9, 9.10, and equation (9.9)]. We will refer to drops capable of achieving such angles as *liquid pearls*.

There are several ways to realize such objects. One is to create highly disordered surfaces similar to those made by the Kao group in Japan.[8,9] These surfaces are made of a layer of hydrophobic spherules (derived from Teflon) with a characteristic size of 1 μm. The combined effects of hydrophobicity, roughness, and trapping leads to (advancing) contact angles of the order of 175° (Figure 9.11). The hysteresis on such surfaces is small (of the order of 10°).

Conversely, we can use a material with a low roughness, but whose design ensures the pinning of lines of contact, thus resulting in super-hydrophobicity.[5] Figure 9.12 is a scanning electron microscopy picture of a surface manufactured by a molding process, whose pillared structure is

FIGURE 9.11. Millimeter-size water drop on a rough, hydrophobic surface. The drop forms a
pearl, with a measured contact angle of 174°. (From T. Onda, S.
Shibuichi, N. Satoh, and K. Tsujii,
in *Langmuir, 12*, p. 2125 (1996).
© 2001 American Chemical Society. Reproduced by permission.)

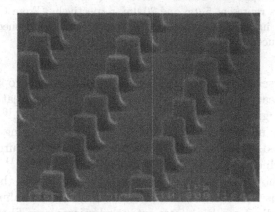

FIGURE 9.12. Ideal textured surface obtained by a molding process. (From J.
Bico, D. Marzolin, and D. Quéré, in *Europhysics Letters, 47*, p. 220 (1999).
© EDP Sciences. Reproduced by permission.)

reminiscent of the drawings that served as the basis for our previous simplified discussion (Figures 9.3 and 9.5).

If the pillars are made hydrophobic, the wetting of such a surface is practically zero (the advancing contact angle is 170°). This we can understand
if we assume that the drops rest on the top of the pillars only (a behavior
that might be called the "fakir effect"). The problem below will highlight
the limitations of such a surface because of the possibility for drops to
become impaled onto the pillars. That is in fact precisely what can be observed with such a system. If one presses down on a drop, the contact angle
suddenly decreases as a result of the impalement. At that point, the drop
no longer benefits from the surface roughness promoting hydrophobicity.
Since the roughness is low, Wenzel's law indicates that the effect on the
contact angle is itself small.

Problem: Show that the fakir state is metastable on a surface such as
in Fig. 9.12, in most cases. [For that, compare the interfacial energy of
the Wenzel state and that of the Cassie state.].[7]

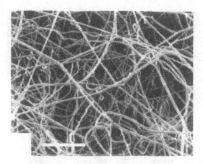 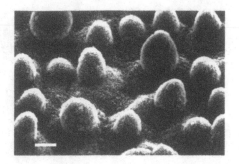

FIGURE 9.13. Naturally occurring super-hydrophobic surfaces, as seen in an electron microscope. Shown are a drosera leaf (left) and a lotus leaf (right). The horizontal bars represent 5 μm and 20 μm, respectively. (From C. Neinhuis and W. Barthlott, in *Annals of Botany, 79*, p. 667 (1997), published by Academic Press. Reproduced by permission.)

Nature herself has created a vast panoply of super-hydrophobic surfaces (blades of wheat, pea pods, lotus leaves, water lilies, duck feathers, butterfly wings), leveraging a mix of hydrophobicity and textures.[10] The textures she favors are most often intermediate between the limiting cases we have just considered. They are generally regular and often feature two scales of roughness, which enhances their hydrophobic properties.[11] As examples, Figure 9.13 shows the surface of a carnivorous plant called the *drosera*, and that of a lotus leaf. In the case of the drosera, we are dealing with a kind of porous medium resulting from entangled microscopic fibers (the bar on the photographs measures 2 μm). In the case of the lotus leaf, we have a rough surface with two roughness pitches (approximately 10 μm and 1 μm, respectively).

There is another way to make liquid pearls, namely, texturing the surface of the liquid. Figure 9.14 shows the object formed when one mixes a hydrophobic powder (such as clay) with a small amount of water. The powder attaches itself to the surface of the drop, forming a kind of shell encapsulating the liquid. No matter what surface such an object is placed on (in the present case, it is glass, which water would normally wet), it displays zero wetting. We call such objects *liquid marbles*.[12]

Finally, there is a third category of pearls—the so-called Leidenfrost drops. The method to create them involves placing a water drop on a very hot plate (typically 300°C). The drop will retain a spherical shape because of the vapor film that supports it. Needless to say, the drop evaporates, but because the vapor film is a good thermal insulator, the evaporation is slow (of the order of a minute for a millimeter-size drop). This phenomenon has been known for a long time (Leidenfrost, 1756), and has been discussed by Bouasse under the name *spheroidal state*. Nevertheless, many questions persist (such as the thickness of the underlying film and the lifetime of the drop).

FIGURE 9.14. Liquid marble obtained by mixing clay (a super-hydrophobic powder) and water. The resulting object, a millimeter in size in the present case, can subsequently be transferred onto a substrate (here glass) that would normally be wetted by water, were it not for the protection of the powder shell. (From P. Aussillous and D. Quéré, in *Nature, 411*, p. 924 (2001). © 2001 McMillan Magazines, Ltd. Reproduced by permission.)

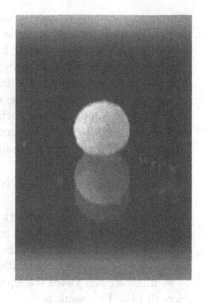

9.2.3.2   Static States

Even in the case of zero wetting (contact angle of 180°), a liquid marble (or a pearl) is not perfectly spherical. Indeed, it gains a small amount of energy by lowering its center of gravity, which implies a certain deformation of its base. Zero wetting notwithstanding, a contact surface forms with the solid on which it rests, as depicted in Figure 9.15 (Mahavedan-Pomeau model).[13] What we refer to as the "contact" is the zone at the bottom of the drop which is parallel to the underlying substrate.

Let δ represent the distance by which the center of gravity is lowered and let l be the size of the contact area that develops (it has been grossly exaggerated in Figure 9.15). In the limit of small deformations ($\delta \ll l$), we can write the geometrical relation

$$l^2 \approx 2R\delta. \qquad (9.13)$$

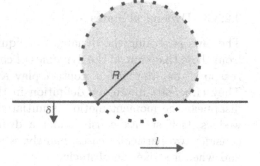

FIGURE 9.15. Gravity-induced contact in zero-wetting regime between a drop and its support acting as an air cushion.

The energy associated with a sagging drop includes a gravity term and a surface tension term of opposite sign, since the surface area increases as a result of the deformation. The increase in surface area is due primarily to the fact that the volume initially in the dotted region shown on Figure 9.15 has had to be redistributed within the drop, resulting in a slight increase in radius. This volume is of the order of $l^2\delta$, causing an increase in radius $\Delta R \propto l^2\delta/R^2$, and hence of surface area $R\Delta R \propto \delta^2$ [where we have made use of equation (9.13)]. In the end, the sagging of the drop implies a change in energy ΔE given dimensionally by

$$\Delta E \propto -\rho g R^3 \delta + \gamma \delta^2 \qquad (9.14)$$

where the first term is the work done by the weight over the sagging distance δ, and the second term is the change in surface energy due to the deformation, assuming that the surface tension below the drop remains equal to γ—a good approximation for porous solids.

Minimizing ΔE yields the equilibrium values of δ [and, hence, l via equation (9.13)]. The result is[13]

$$\delta \propto R^3 \kappa^2 \qquad \text{and} \qquad l \propto R^2 \kappa \qquad (9.15)$$

where κ^{-1} is the capillary length (equal to 2.7 mm for water). Equation (9.15) holds, of course, as long as the drops are spherical, i.e., when R is small compared to κ^{-1} (larger drops are flattened by gravity and become puddles). For a millimeter size drop, we find that l is approximately 100 μm and δ is of the order of 20 μm.

Finally, we note that this calculation assumes that wetting is strictly zero. In the case of partial wetting [equation (9.9), Figure 9.12], and if ε denotes the small deviation of the contact angle from π, a contact zone forms due to wetting. Its magnitude in zero gravity is $l^* \approx R\varepsilon$. Drops whose size R is less than $\varepsilon\kappa^{-1}$ will obey this last relation, whereas equation (9.15) will apply to drops larger than this value (but smaller than κ^{-1}), for which gravity imposes a contact zone larger than dictated by capillary forces alone ($l > l^*$).

9.2.3.3 Dynamical States

The laws governing the dynamics of liquid pearls are distinctly different from those that control the dynamics of conventional wetting. The primary reason is that the lines of contact play a negligible role for liquid pearls. They either are absent by definition in the case of zero wetting, or they disappear the moment motion is initiated because of the film of air that wedges itself spontaneously under a drop in a dynamical situation. We consider two particular cases, namely, when a drop slides down an incline and when it strikes an obstacle.

Descent Down an Incline

The first remarkable observation is that a liquid pearl comes down an incline with extreme ease even if there is some minor contact due to wetting. We have already emphasized that the hysteresis in the case of quasi-zero wetting is exceedingly small (Figure 9.1) because of the apparent homogeneity of the supporting surface, which turns out to be made primarily of air. In contrast to what happens ordinarily when submillimeter size drops remain stuck to substrates because of hysteresis, a drop in a quasi-zero-wetting situation generally has no difficulty sliding down an incline.

> **Problem:** Calculate the pinning threshold for a drop of radius R on an inclined plane tilted by an angle α relative to the horizontal direction in a quasi-zero wetting situation ($\theta^* = \pi - \varepsilon$ and $\Delta\theta \ll \theta^*$, where $\Delta\theta$ denotes the difference between advancing and receding angles). Hint: Write down that that the trailing half of the drop joins the solid with a receding angle $\theta^* - \Delta\theta/2$ and that the leading half of the drop does likewise with an advancing angle $\theta^* + \Delta\theta/2$. Expand the relevant mathematical expressions in a power series in ε and show that the pinning condition is critical with respect to ε (meaning that it increases rapidly as ε decreases).

When the drops are viscous, they display an odd behavior as they slide down at constant velocity. The smaller pearls turn out to descend faster than the larger ones! This can be traced back to the non-linear law governing the size of the contact vs. that of the drop, as Mahadevan and Pomeau have shown.[13] A viscous drop (Reynolds number smaller than 1) will pick whatever descent mode minimizes its viscous dissipation. That mode happens to be a *solid rotation* (much as in the case of a steel ball), in which instance the dissipation within the drop vanishes altogether. The only friction to consider is then that which develops at the contact itself. If the drop comes down at a velocity V, the associated velocity gradient is V/R^* and the viscous force F_V is $(\eta V/R)l^2$, where we have integrated the viscous stress over the area of the contact. The weight W in the reduced gravity imposed by the inclined plane is $\rho R^3 g \sin\alpha$. The steady-state descent of the pearl is obtained by writing the equilibrium of the pertinent torques since the drop is in overall rotation, which can be written as $WR = F_V l$. We deduce the law of descent by eliminating l via equation (9.15). The result is[13]

$$V \propto \frac{\gamma}{\eta}\frac{\kappa^{-1}}{R}\sin\alpha \qquad (9.16)$$

---

*In the contact zone of size l, the velocity varies from 0 (near the solid) to Vl/R (at the edge of the contact zone).

This law is consistent with the paradoxical observation that small pearls descend faster than large ones. Since κ^{-1} decreases with increasing g (as $1/\sqrt{g}$), the velocity does also decrease as g increases (!). A higher gravity imposes a wider contact, enhancing friction effects.

For equation (9.16) to be valid, the drop must retain its static shape [equation (9.15)] as it comes down. However, both viscosity and inertia could induce shape changes apt to affect the dynamics of the motion. To use equation 9.16 is therefore necessary to be in a regime where surface tension effects (which make the drop spherical) predominate. Recall that the capillary number $Ca = \eta V/\gamma$ compares viscous and capillary effects, while the Weber number $We = \rho V^2 R/\gamma$ compares inertial and capillary effects. The requirement is that both numbers be small compared to 1, which in either case mandates that the velocity be low.

> **Problem:** If the size of the drop becomes comparable to κ^{-1}, equation (9.16) predicts that the descent velocity is $V \propto \gamma/\eta \sin \alpha$. Drops of that size (or larger) are flattened into puddles, as discussed in chapter 2. Prove the above velocity relation for a puddle by writing the equilibrium between the force of gravity and the viscous friction force. (It is worth noting that this velocity does not depend on the gravitational force ρg.)

Deformations: Peanuts and Liquid Wheels

When the liquid is less viscous or the slope higher, descent can occur at higher velocities. The rotation imparts to the liquid a centrifugation energy, which causes rather peculiar changes in shape. The drop, which started out spherical, can deform into a disk, even a peanut (Figure 9.16a) or a torus (Figure 9.16b).[14]

Consider the case of a torus. If we wish to establish a criterion for the deformation of a spherical drop, we must compare inertia (the centrifugal force being responsible for distorting the drop) and capillarity (which wants to preserve a spherical shape). As stated above, the parameter that

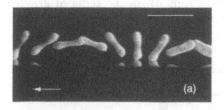

FIGURE 9.16. Time-lapse photographs of a liquid marble descending a slope at a velocity of 1 m/s. The camera is tilted to the same inclination as the solid and the movement takes place toward the left (the bar indicates a distance of 1 cm). Because of the centrifugal force, the sphere deforms into a peanut (a) or a torus (b). (From P. Aussillous and D. Quéré, in *Nature, 411*, p. 924 (2001). © 2001 McMillan Magazines, Ltd. Reproduced by permission.)

FIGURE 9.17. Liquid torus descending a slope.

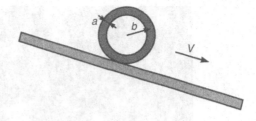

compares these two quantities is the Weber number We, given by

$$We = \frac{\rho V^2 R}{\gamma} \qquad (9.17)$$

Deformation occurs when We is no longer negligible compared to 1. Next, we can evaluate the characteristics of the liquid torus (Figure 9.17) by writing the equilibrium between the centrifugal force (per unit volume) and the gradient of the Laplace pressure, which tends to reduce the radius of the torus.

If a and b denote the two radii of the torus (we assume $a \ll b$ for simplicity), the equilibrium (written dimensionally) reads

$$\frac{\rho V^2}{b} \propto \frac{\gamma}{ba}. \qquad (9.18)$$

By using the conservation of volume ($R^3 \propto ba^2$, where R denotes the radius of the parent drop), we can deduce the size of the liquid wheel:

$$b \propto R \cdot We^2. \qquad (9.19)$$

This result is sensible in that b indeed becomes larger than R when the Weber number becomes larger than 1. This law was proposed (in a somewhat less explicit form) by Rayleigh to explain one of Plateau's experiments, in which he placed a drop of oil on a thread and set it spinning around its axis, whereupon he noticed the formation of annular structures.[14] There exists a sophisticated version of Rayleigh's calculation due to Chandrasekhar, who discussed the stability of these objects.[15] This great astrophysicist (famous for inventing black holes) was, incidentally, intrigued by the similarity of this problem (*in appearance*, as he himself put it) with some celestial objects such as Saturn's rings.

Impacts and Bounces

One of the most fascinating properties of super-hydrophobic substrates is their behavior upon impact of a liquid drop.[16] Figure 9.18 shows the trajectory of a drop of water launched from a height of 1 cm above a substrate that is slightly inclined (at an angle of 1°).

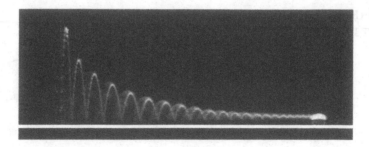

FIGURE 9.18. Trajectory of a 1-mm water drop impacting a slightly inclined super-hydrophobic substrate. (From D. Richard and D. Quéré, in *Europhysics Letters, 50*, p. 769 (2000). © EDP Sciences. Reproduced by permission.)

Many bounces on the surface are clearly visible. This is a direct consequence of non-wetting. As soon as it hits the solid, the drop deforms while retaining an angle of the order of 180° with respect to the solid. This process enables the drop to store its kinetic energy as surface energy while minimizing losses by viscosity, which are normally responsible for dissipating the energy of a drop upon impact (precluding rebounds) and ordinarily slow it to a stop.

Here, instead, the drop *rebounds*—a particularly interesting phenomenon from a practical point of view. Still, Figure 9.18 shows a noticeable dampening of the bounces. This is primarily related to the liquid nature of the spring. At the moment of impact, the solid induces a velocity gradient within the drop. Part of the kinetic energy is transformed into vibrational energy within the drop. Every time the drop takes off, its vibration is evidenced in the figure by the wavy appearance of the luminous streak. The viscosity of the liquid attenuates this vibration during the course of successive impacts: this explains the damping of the rebounds.

The drop shown in Figure 9.18 deforms relatively weakly during impact; this allows it to rebound as a whole. To have weak deformations, the kinetic energy must remain small compared to the surface energy; this implies a Weber number [equation (9.17)] smaller than 1. This condition is undoubtedly met in Figure 9.18, where We starts out below unity and becomes ever smaller as rebounds accumulate. For more violent impacts, the deformations are much more pronounced and the drop ejects droplets. Figure 9.19 shows the rebound of a 1-mm drop for which the Weber number is equal to 16. Note the formation of an object shaped somewhat like a baseball bat. The droplets are ejected at the top of the bat. The vibrations of the bat following the ejection of droplets spawn very bizarre shapes.

All these systems are at once capillary and inertial, and it is easy to deduce a few of their characteristics. A natural question to ask concerns the contact duration upon such impacts. It was first raised by Hertz in the case of solid/solid collisions accompanied by rebounds.[17] Experiments

FIGURE 9.19. Take off of a water drop following a violent impact on a super-hydrophobic surface. The drop experiences fragmentation and highly non-linear oscillations (courtesy Denis Richard and Christophe Clanet).

with a liquid marble show that, over a large range of impact velocities (from 20 to 200 cm/s), the contact duration is independent of the velocity.[18] It does increase with the radius of the drop as $R^{3/2}$. This dependence is characteristic of a behavior where capillarity and inertia are the primary players. Indeed, if we write the dynamical equation by including only these two contributing terms, the result can be written dimensionally as

$$\rho \frac{R}{\tau^2} \propto \frac{\gamma}{R^2} \qquad (9.20)$$

which yields an impact duration of the form

$$\tau \propto \left(\frac{\rho R^3}{\gamma} \right)^{1/2}. \qquad (9.21)$$

This last result is independent of the velocity and agrees with the experimentally observed dependence on the radius. For a 1-mm drop, the time is of the order of a few milliseconds. The time τ scales as the characteristic oscillation period of a drop vibrating in air. The dynamics of the oscillations is described by writing the equilibrium between capillarity and inertia. The calculations were done by Lord Rayleigh.[19] He worked out the appropriate numerical coefficients applicable to the various vibration modes of drops—an issue already touched upon in section 5.2.

9.3 Wetting and Porous Media

9.3.1 Capillary Rise in a Porous Medium

Imagine that we take a piece of blotting paper and dip its end into a cup of coffee. The coffee will promptly ascend (Figure 9.20) because the blotter has a lower surface energy when wet. In our technical jargon, we express this by saying that $\gamma_{SO} > \gamma_{SL}$.

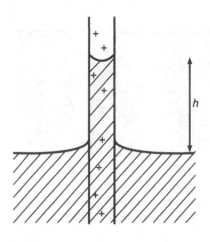

FIGURE 9.20. Capillary rise in a porous material (indicated by the crosses).

The rise of the liquid will eventually stop at a certain height h. The capillary forces pull the liquid up, whereas gravity pulls it down. Our objective is to determine h. More specifically, we consider the following situation: (a) $\gamma_{SO} > \gamma_{SL}$, which means that the blotting paper soaks up the liquid. (b) The equilibrium contact angle θ_E of the porous material is finite ($\theta_E > 0$). Each pore is in a partial wetting regime. This ensures that at an altitude $z < h$ the pores are completely filled, whereas they are empty for $z > h$.*

The problem is quite similar to the one we considered in the context of Jurin's law (chapter 2). But instead of vertical capillaries with a simple cross section, we are dealing here with a geometrically more complex medium made of random pores. Nevertheless, it is possible to find h by means of a macroscopic argument.

A fundamental characteristic of a porous medium is its specific surface area Σ (expressed in m^2 per kilogram of material). Qualitatively, we have $\Sigma \cong (\rho_S d)^{-1}$, where ρ_S is the density of the compacted solid devoid of pores (typically, $\rho_S \approx 1$ g/cm^3), and d is the diameter of the capillary. For a pore diameter $d = 10$ µm, Σ is of the order of 100 m^2/kg. Another important parameter of the porous medium is its void fractional volume Φ. Therefore, its average density is $\rho_S(1 - \Phi)$. The surface area Σ_V per unit volume is

$$\Sigma_V = \Sigma\rho_S(1 - \Phi) \cong 1/d. \tag{9.22}$$

Consider now a column made of this material, with cross-sectional area S and a wet volume Sh (Figure 9.20). When the height increases from h to $h + dh$, there is a corresponding change in capillary energy given by

$$dE_{cap} = \Sigma_V S\, dh(\gamma_{SL} - \gamma_{SO}). \tag{9.23}$$

---

*In the opposite case (total wetting), the possibility exist to form films thinner than the dimension of the pores (thickness comparable to the pancake thickness discussed in chapter 4). In such a case, we would have to distinguish 3 regions in the blotting paper, namely, completely wet, partially wet, and dry.

In the process, we have had to raise a mass dM of liquid (assuming that all the pores making up the void fraction Φ are interconnected and ready to be filled):

$$dM = \rho_L \Phi S \, dh \tag{9.24}$$

where ρ_L is the density of the liquid. Associated with dM is a change in gravitational energy dE_g:

$$dE_g = gh \, dM. \tag{9.25}$$

At equilibrium, the height h must be such that $dE_{cap} + dE_g = 0$, from which we deduce

$$h = \frac{\Sigma_V(\gamma_{SO} - \gamma_{SL})}{\rho_L \Phi g}. \tag{9.26}$$

This expression for h is completely analogous to Jurin's law, but it has the advantage of involving explicit macroscopic parameters of the porous medium. Qualitatively, for $\Phi \approx 0.5$, we have

$$h \approx \frac{\kappa^{-2}}{d} \cos \theta_E$$

When $d = 10$ μm, the result comes out to $h = 10$ cm.

9.3.2 Equilibrium Angle at the Surface of a Porous Medium

Is there a macroscopic contact angle between a liquid and a porous medium? The answer is that an average contact angle $\bar{\theta}_E$ can be defined only in an equilibrium situation. To illustrate the point, when $\gamma_{SL} < \gamma_{SO}$, it would make no sense to consider the contact between a dry porous material and a liquid since the liquid would instantly and irreversibly penetrate the porous medium.

Figure 9.21 depicts two cases where the liquid interacts with a saturated porous medium. Case (a) consists of a vertical porous plate exposed to

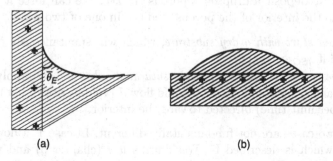

FIGURE 9.21. (a) Meniscus alongside a wet porous material; (b) drop resting on a wet porous surface.

a liquid pool, and the two connect by way of a meniscus of angle $\bar{\theta}_E$. In case (b), a horizontal porous substrate has been prepared ahead of time and a drop is subsequently deposited on top of it. Case (a) is more clearly defined than case (b) because saturation is achieved via a *unique* pathway, namely, capillary rise from the pool. Case (b) is a bit trickier because the outcome depends on the precise way the wet substrate was prepared. If prewetting was realized by spreading the liquid onto the surface, there is always the danger that either too much or too little liquid was used. The best substrate preparation method in case (b) is probably also by way of capillary rise, by dipping it upside down into a liquid pool.

Cautionary Remarks

- The reader should bear in mind that the definition of $\bar{\theta}_E$ remains an approximation for several reasons. Among them are the following:

  1. Hysteresis can occur locally within the pores;
  2. Air bubbles can be trapped during capillary rise.

- The vertical rise discussed here is the most common outcome. However, other interesting situations can arise, which have been discussed by Marmur.[20] For example, in the absence of gravity effects, a drop placed on a dry, horizontal, porous material (again, with $\gamma_{SO} > \gamma_{SL}$) may enter and occupy a restricted region of the substrate.

That said, it is possible to define and measure an angle $\bar{\theta}_E$. There are, however, cases where one may end up in a total wetting situation.

An unavoidable complication has to do with the fact that porous media always have a somewhat random surface. As a result, hysteresis effects often show up. It is not uncommon, for instance, to encounter a receding angle that is zero. This indicates that the entire periphery of a drop placed on a surface (Figure 9.21b) is pinned on the surface.

9.3.3 Suction Experiments on Drops

Suppose we deposit a drop on a porous surface. We can force it to drain itself into the interior of the porous medium in one of two ways:

1. *We can start with a dry substrate*, which will spontaneously draw the liquid if $\gamma_{SO} > \gamma_{SL}$;
2. *Or we can start with an already saturated substrate* and apply to it a mechanical suction inducing a liquid flow J (volume drawn per unit area and per unit time) directed toward the interior.

The two cases are not fundamentally different. In case 1, a flow $J(t)$ is set up, which is described by Washburn's law (chapter 5) and varies as $t^{-1/2}$. In case 2, The flow J is independent of time, which simplifies things. In this section, we shall discuss primarily case 2.

Consider a situation where we subject the drop depicted in Figure 9.21b to a constant flow J, starting at time $t = 0$. We assume the presence of a certain hysteresis, as is often the case, and we chose a drop whose size is less than κ^{-1}, which ensures that gravity effects are negligible.

- *At the very beginning of the suction*, for the first few milliseconds, the suction creates an underpressure near the edges of the liquid, and the drop ceases to be a spherical cap.[21] This initial regime is practically unobservable because it lasts an extremely short time and it involves only small changes in slope. It has, however, been studied theoretically, at least in the limit of small contact angles θ. The conclusion is that wavelets propagate from the edges with a velocity $V_S \approx V^* \theta^3$.
- *As soon as the time reaches a value* such that $t > r/V_S$ (where r is the horizontal radius of the drop, which remains constant as long as $\theta > \theta_r$), the pressure becomes practically constant again throughout the drop. According to Laplace, the curvature of the profile must be constant as well. We then recover a spherical cap, albeit with an angle $\theta(t)$ less than the initial angle $\bar{\theta}_E$. We can in fact relate $\theta(t)$ to the volume of the drop. At time t, the volume is

$$\Omega(t) = \Omega(0) - \pi r^2 J t. \tag{9.27}$$

It is also the volume of a spherical cap

$$\Omega(t) = \frac{\pi}{4} r^3 \theta(t) \quad \text{(with } \theta \ll 1\text{)}. \tag{9.28}$$

Combining equations (9.27) and (9.28) leads to a linear decrease of $\theta(t)$ in time, which has recently been verified experimentally.[22]
- *If there is a receding angle $\bar{\theta}_r > 0$*, a time comes when $\theta(t) = \bar{\theta}_r$, at which point the triple line breaks free and the drop retracts.[22]

9.3.4 Suction Experiments on Films

We consider here a porous ribbon emerging vertically from a liquid pool at a certain velocity V (Figure 9.22). The porous medium is wetted on its surface but not in its interior, and a certain penetration flow J is set up. We assume a situation where the velocity V is sufficiently high for a Landau-Levich film to be dragged (with a certain thickness e discussed in chapter 5). But the film cannot retain a constant thickness since it is being extracted toward the interior of the ribbon at a rate J. Therefore, it survives only up to a height h. What is the value of this height?

The theoretical answer is quite simple if J can indeed be considered constant throughout the film. The velocity field is no longer strictly vertical, but has a horizontal component precisely equal to J. This results in a triangular profile (Figure 9.22) and a height $h = eV/J$.

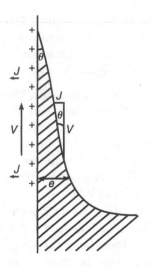

FIGURE 9.22. Film drawn by a porous support.

The astute reader may well be startled by the assertion that the dynamical contact angle is $\theta = J/V$. As we know, the dynamical contact angle is usually determined by dissipation mechanisms in the vicinity of the triple line, and there is no reason why it should be equal to J/V. Nevertheless, detailed theoretical analyses suggest that the distortions of the profile associated with the line are restricted to a microscopic region near the line.[23] The macroscopic angle is indeed $\theta = J/V$.

9.4 Wetting at Soft Interfaces

The stability of a liquid film wedged between a hard solid S (e.g., a metal or glass) and a soft material R (e.g., rubber) is of critical importance in numerous applications. If the liquid is a lubricant (or tears), any rupture of the liquid film can cause serious damage (friction and failure of metal parts, injury to cornea, etc.). Conversely, when you are driving on a wet road, the film of water between the tires and the pavement must be eliminated quickly if there is any hope of retaining control of your vehicle. Depending on the wetting properties of the sandwiched liquid, the ultimate outcome will range from strong adhesion to free sliding of the soft solid on the rigid substrate. We will start out by describing the wetting characteristics of a liquid at the S/R interface, and then move on to discussing the stability of wedged-in films. The fundamental physical constant is the spreading parameter S characteristic of the liquid trapped by the elastomer.

- If $S > 0$, the liquid wets the contact totally The interface always remains lubricated in the presence of a thin wetting film that is stable. Such a situation has been thoroughly studied, notably by A. Roberts.[24]

- If $S < 0$, wetting is partial:

  1. A drop trapped at the interface does not spread. It does not form a spherical cap, but is flattened because it must deform the soft material surrounding it. Its shape, which will enable us to measure S, will result from competition between surface and elastic energies.
  2. A wedged-in film is metastable. We will describe the phenomenon of dewetting by nucleation and growth of a dry contact.

After dewetting has taken place, one obtains a dry adhesive contact between the elastomer and the rigid substrate. We will see that if one shears such a contact by displacing the solid at an increasing velocity U, the contact will be invaded—first partially and then completely—by the liquid. The loss of contact explains the perilous phenomenon of aquaplaning when a driver brakes on a wet roadway, causing the driver to lose steering control.

A great deal of research has been devoted to the subjects of friction, lubrication, and wear out since D. Tabor's early work. Generally, one relies on liquids that have the ability to wet the contact $(S > 0)$. Such liquids are referred to as lubricants. We will not debate these, and the interested reader is encouraged to consult a book on lubrication.[25]

We are interested here in the case of non-wetting liquids. The dynamics of dewetting, that is to say, the establishment of adhesive contacts, will define the "setting time" required for a soft object to stick onto a solid substrate. This has numerous applications, including the adhesion of tires, the setting of glues on wet media (such as boats), as well as the vast field of adhesion of living cells in biology. We will illustrate this particular issue by describing the adhesion of mushroom spores on a support— a crucial step controlling their development. To simulate the surface of rice plants, we may select Teflon, which has a comparable degree of hydrophobicity.

9.4.1 Principles of "Elastic" Wetting

The physics of drops placed on a substrate in open air is well understood. What happens if air is replaced by a soft solid?

We focus on the case of a drop of liquid L sandwiched between a rigid solid S of low energy (e.g., silanized glass) and a semi-infinite rubber R. We assume the elastomer to be homogeneous and isotropic. This material is "soft" in the sense that it can easily be deformed. If we apply a stress σ, the resulting deformation ε is given by Hooke's law:[26]

$$\sigma = E\varepsilon \qquad (9.29)$$

where E is the elastic modulus, of the order of 1 MPA. By way of comparison, the elastic modulus of a rigid material is of the order of 10^5 MPA.

9.4.1.1   The Spreading Parameter S

A wedged-in drop will either spread or remain collected depending on the sign of the spreading parameter defined by

$$S = \gamma_{SR} - (\gamma_{SL} + \gamma_{LR}) \tag{9.30}$$

where γ_{ij} represents the surface tension at the S/R, S/L, and L/R interfaces, respectively.

- If $S > 0$, the drop spreads into a nanometer-size film.
- If $S < 0$, the drop does not spread. If it were exposed to air, it would form a spherical cap with a contact angle θ_E given by Young's relation (chapter 1). A simple measurement of θ_E would determine S. Unfortunately, when the liquid is encapsulated in a soft matrix, this method is unusable.

9.4.1.2   Young's Relation No Longer Holds!

Young's relation is valid only in the immediate vicinity of the line since the shape of the drop will be determined by the elastic deformation of the elastomer. Together with the spreading parameter S and Young's modulus E of the rigid substrate, we can introduce a characteristic length h_0, which we call the *elastic length*, defined by

$$h_0 \equiv \frac{|S|}{E} \tag{9.31}$$

With $S \approx 10$ mN/m and $E \approx 10^6$ Pa, we get $h_0 \approx 10$ nm.

While the capillary length κ^{-1} (introduced in chapter 1) describes the competition between gravity and capillarity, the length h_0 describes the competition between the deformation energy of the soft matrix and the surface energy. Young's relation would be valid only over dimensions smaller than h_0, where other types of forces (such as van der Waals) modify the profile anyway.

We will see shortly that analyzing the shape of a liquid drop trapped at the solid/elastomer interface enables us to determine the value of S.

9.4.1.3   Penny-Shaped Trapped Drops

Figure 9.23 depicts a drop trapped between a rigid solid and rubber.

Omitting all numerical coefficients, the energy of the drop can be written as

$$G(H, R) \propto -SR^2 + E \left(\frac{H}{R}\right)^2 R^3 \tag{9.32}$$

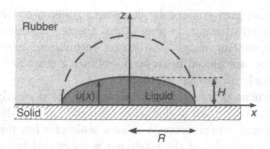

FIGURE 9.23. Penny shape of a trapped drop.

where H is the height of the drop at its center, R^2 is its surface area, and E is the elastic modulus of the elastomer.

The first term in equation (9.32) is the interface contribution corresponding to the surface energy that must be expended to form a wet zone of radius R, starting from a dry rubber/solid interface. The second term is the elastic energy associated with a typical deformation (H/R) induced by the drop in the elastomer. The deformation extends over a depth of the order of the radius R of the drop (the deformation is localized within a volume $\approx R^3$).

Minimizing the energy while keeping the volume Ω constant ($\Omega \propto HR^2$) yields

$$H^2 \propto h_0 R. \tag{9.33}$$

For $R = 1\,\mu m$ and $h_0 = 10$ nm, we get $H = 100$ nm. The drop is therefore quite flat, resembling a penny. If we manage to measure the thickness H and radius R of the wedged-in drop, we can deduce the spreading parameter S by way of the relation

$$-S = \frac{\pi}{6} E \frac{H^2}{R}. \tag{9.34}$$

The prefactor in equation (9.36) is taken from a paper by I. Sneddon, who calculated the energy [equation (9.32)] exactly.[27]

9.4.2 Experimental Observation of Elastic Wetting

In the previous section, we have established the wetting laws at soft interfaces. We now proceed to describe an experiment on elastic wetting performed by P. Martin.[28, 29]

9.4.2.1 The Three Partners: Soft Solid, Liquid, and Elastomer

- *The elastomer* consists of cross-linked PDMS (polydimethylsiloxanes), synthesized from two liquid components, molten PDMS and the cross-linking agent.

PDMS lenses are obtained by depositing millimeter-size drops of the non-cross-linked mixture on glass slides treated chemically to prevent PDMS from wetting them. The lenses, forming an angle of 70° with the glass slide, are placed in an oven. Following cross-linking, the lenses detach themselves from the glass slides.

The specimens obtained by this technique are perfectly transparent. After baking in air, their surfaces are optically smooth. No roughness is detectable under a microscope, i.e., on a scale of a few microns.

The Young modulus of the elastomer is measured by straightforward compression of a cylindrical sample. Its value is $E = 0.74 \pm 0.05$ MPA. The elastomer, affixed to a mechanical arm, is linked to a micromanipulator providing motion in three directions.

• *The substrate* is a silanized microscope slide. The particular silane used to prepare the slide is octadecyltrichlorosilane (OTS), whose formula is $CH_3-(CH_2)_{17}-SiCl_3$. It is the most commonly used trichlorosilane for this type of reaction. The resulting hydrogenated surfaces have a quality slightly lower than is afforded by the almost ideal surfaces of silicon wafers used in lieu of microscope slides. Indeed, glass slides are rougher than the typical surface of silicon wafers (~ 5 Å) and are chemically more heterogeneous as well (various ions "pollute" the top layer of silica), but they are transparent. The slides are mounted on a two-axis motorized translation stage.

• *The liquid* is a fluorinated oil called PFAS (polyfluoromethylalkylsiloxane). Since the dewetting kinetics of water is far too rapid to be observable with a regular camera, PFAS is a much better choice. It dewets much more slowly and does not dissolve PDMS. The number of monomers of this fluoro-silicone determines its viscosity, which is 100 to 1,000 times higher than that of water. Experiments with a fast camera show that water obeys the same laws.

9.4.2.2   Observation of the Contact: Reflection Interference Contrast Microscopy

The experimental setup relies on a reflection interference contrast microscope (RICM) used in conjunction with a CCD camera and a video camera recorder (VCR).

The specimens are observed under monochromatic illumination. The profile $u(x)$ of the liquid/rubber interface can be deduced from the interference pattern generated by rays reflected off the glass/liquid interface (refractive indices n_0 and n_1, respectively) on the one hand, and off the liquid/rubber interface (index n_2) on the other (Figure 9.24).

The samples are placed on the stage of an inverted microscope. Illumination is provided by a mercury arc lamp passed through a band pass filter centered at $\lambda = 546$ nm ($\Delta\lambda = 10$ nm) to select the green line. The illumination is episcopic, meaning that it goes through the microscope objective.

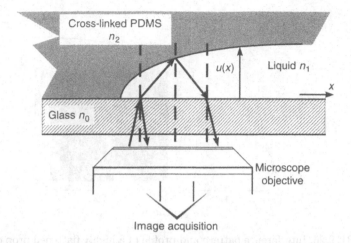

FIGURE 9.24. Principle of the reflection interference contrast microscopy (RICM) technique.

This arrangement obviates the need for direct lighting and collects only the light reflected at the interfaces of interest.

9.4.2.3 Drop Profile and Measurement of S

How does one trap drops at the glass/rubber interface?

- In the case of water, the answer is by injection with a syringe (Figure 9.25a). A fairly high pressure is required because the elastomer sticks naturally to glass and it is difficult to break the adhesive contact.

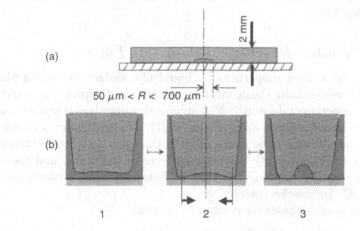

FIGURE 9.25. Deposition of drops at a glass/elastomer interface by (a) injection (water) or (b) dewetting (PFAS). A film (1) squeezed between a rubber marble and glass thins down (2) and dewets (3). Note the dimple formed during drainage (2).

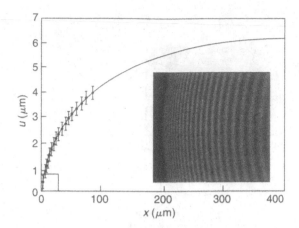

FIGURE 9.26. Interference pattern and profile of a highly flattened drop of water sandwiched between an elastomer and glass. Note the different scales on the two axes. (From P. Martin, P. Silberzan, and F. Brochard, in *Langmuir, 18*, p. 4910 (1997). © 2001 American Chemical Society. Reproduced by permission.)

- For high-viscosity PFAS, injection is no longer an option. Instead, one can take advantage of the spontaneous dewetting of a PFAS film sandwiched between a PDMS lens and glass (Figure 9.25b).

Figure 9.26 shows the profile of a drop, as determined by RICM. By changing the volume of the drops, it is possible to construct a curve displaying H^2 as a function of the radius R, which makes it possible to deduce h_0 by means of equation (9.33). If the elastic modulus is known, S can be deduced from equation (9.31). The results are $h_0 = 64 \pm 5$ nm (hence $|S| = 50 \pm 5$ mN/m) for water, and $h_0 = 10 \pm 5$ nm (hence $|S| = 7 \pm 1$ mN/m) for PFAS.

9.4.3 "Elastic" Dewetting of Wedged-in Films

To study the wetting properties of a liquid film wedged between a planar solid and a semi-infinite elastic medium, the standard approach is to use the following experimental setup.[29,30] A spherical rubber lens is squeezed on a glass slide above a liquid drop (Figure 9.27). The lens deforms elastically and a liquid a few hundred nanometers in thickness finds itself trapped at the interface (Figure 9.28). Since the surfaces of the rubber and the solid are both optically smooth, it is easy to follow the time evolution of the film's profile by interferometry.

Two sequential phases are normally observed:

- A thinning phase of the film by way of drainage;
- a dewetting phase, either starting spontaneously or initiated by a defect etched either on the glass or on the rubber lens.

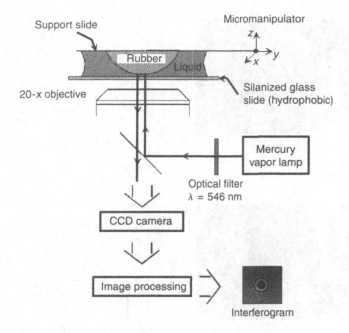

FIGURE 9.27. Schematic of the experimental setup used to trap a film and observe it interferometrically while in the process of dewetting.

9.4.3.1 Drainage

Suppose we press on the lens with a force F. The wedged film will thin down (Figure 9.28b).

If e denotes the thickness of the sandwiched film, V its draining velocity, a the size of the flat portion of the elastomer lens, and η the viscosity of the liquid, we can write the two following equations:

1. The transfer of mechanical energy into viscous dissipation. Dimensionally, the equation reads

$$F\dot{e} \propto \eta \left(\frac{V}{e}\right)^2 e a^2. \qquad (9.35)$$

2. The conservation of liquid volume at the contact:

$$\frac{\dot{e}}{e} = \frac{V}{a}. \qquad (9.36)$$

Combining equations (9.35) and (9.36) yields an expression for the thickness e:

$$e \propto \sqrt{\frac{\eta a^3}{F}} V^{1/2}. \qquad (9.37)$$

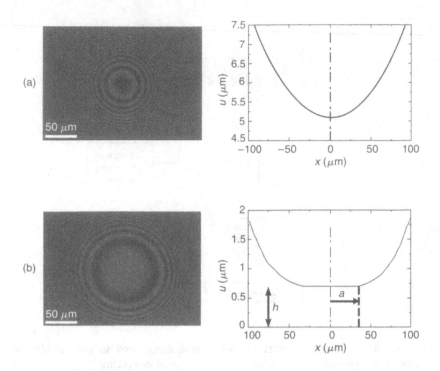

FIGURE 9.28. Approach (a) and squeezing (b) of an elastomer lens, followed by trapping of a wedged film. The profiles at right are deduced from the interferograms at left (courtesy P. Martin).

which, in turn, leads to Reynolds' thinning law:

$$e(t) \propto t^{-1/2} \tag{9.38}$$

This law has been verified experimentally. We should note that the pressure field within the flowing liquid causes the rubber to deform slightly into a dimple (Figure 9.29a).[31] This deformation diminishes as the film thins down. Eventually, the film becomes virtually flat for thicknesses $e \approx 100$ nm. As the film is allowed to drain out, spontaneous dewetting takes place (Figure 9.29). The process starts at the edge of the film because of the presence of the dimple. The film collapses, leaving a few drops trapped at the interface, which can be used to measure S.

9.4.3.2   Controlled Dewetting: Nucleators

In this section, a defect is placed either on a glass substrate or on the elastomer lens.[32, 33] If the defect has a size greater than a critical radius R_c, it will initiate dewetting by nucleation and growth of a dry contact.

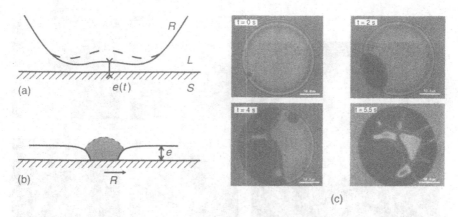

FIGURE 9.29. (a) Drainage; (b) nucleation of a contact; (c) spontaneous dewetting (courtesy P. Martin).

Critical Radius R_c for Nucleation

Upon creating a contact between the elastomer and the solid, the surface energy decreases, but the rubber is deformed. The energy of the contact of radius R in a film of thickness e can be written as

$$G(R, e) \propto S\pi R^2 + 2\pi R E \left(\frac{e}{R}\right)^2 R^3$$

where e/R is the elastic deformation extending over a volume R^3 (Figure 9.29b). The energy $G(R)$ goes through a maximum when the radius R is equal to a critical value R_c given by

$$R_c \propto \frac{e^2}{h_0}. \tag{9.39}$$

For $e = 10$ µm and $h_0 = 10$ nm, we find $R_c = 1$ cm. For $e = 100$ nm, $R_c = 1$ µm. These arguments prove that, if the size of the defects is of the order of a micron, a sandwiched film is metastable and can dewet only at very small thicknesses ($e \approx 100$ nm).

To test the validity of equation (9.39), A. Martin (who succeeded P. Martin in the same laboratory) studied the optimal size and shape of a dewetting "nucleator." He prepared controlled defects of sizes b ranging from 1 µm to 20 µm by means of a Vickers diamond (traditionally used to test the hardness of solids) leaving its imprint on glass surfaces. He was able to show that a sandwiched film, which thins down by drainage, suddenly retracts by dewetting at a thickness $e^* = \sqrt{bh_0}$, deduced from equation (9.39) when $R_c = b$.

Growth of the Contact

P. Martin was the first to observe the growth of a unique contact by placing a defect at the center of the film (Figure 9.30b).[29] As the contact (dark

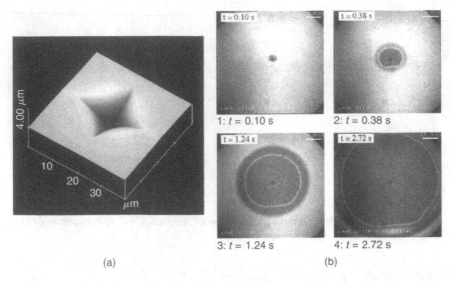

FIGURE 9.30. Dewetting nucleation initiated by a controlled defect on the surface of glass (made by Vickers imprint). (a) AFM image of a defect (courtesy O. du Roure); (b) images of nucleated dewetting (courtesy P. Martin). The white line at upper right represents 20 μm.

zone) expands, we can see the formation of a ridge surrounding the hole and collecting the liquid. The shape of the ridge can be determined by RICM. Its width l is related to its height h via the law $l \propto h^2/h_0$, which shows that the profile of the ridge is quasi-static.

The time evolution $R(t)$ of the contact has been measured for a series of fluorinated oils. The experimental law turned out to be

$$R(t) = k(\eta)t^{0.75\pm0.05} \tag{9.40}$$

where $k(\eta) \propto \eta^{-0.6\pm0.1}$. The width $l(t)$ of the ridge depends on time as

$$l(t) \propto t^{0.4\pm0.1}. \tag{9.41}$$

These laws have been interpreted in terms of a simple dimensional model based on three hypotheses:[34]

1. The ridge has a quasi-static shape. Its width l and its height h are related by the dimensional relation

$$l \propto h^2/h_0. \tag{9.42}$$

2. The liquid missing from the hole is entirely captured in the ridge

$$lh \propto eR. \tag{9.43}$$

3. Viscous dissipation takes place exclusively within the liquid. The elastomer behaves purely elastically. The tally of the surface energy dissipated in friction is written as

$$|S|V = \eta \left(\frac{V}{h}\right)^2 hl. \tag{9.44}$$

The three equations (9.42–9.44) lead to

$$\eta \frac{V}{h_0} (eRh_0)^{1/3} = S. \tag{9.45}$$

If we start out at $R \approx R_c$, the initial velocity $V(e)$ is given by

$$V \propto Sh_0/\eta e. \tag{9.46}$$

Since $V = dR/dt$, integrating equation (9.45) leads to

$$R(t) = k \left(|S|\frac{t}{\eta}\right)^{3/4} \frac{h_0^{1/2}}{e^{1/4}} \tag{9.47}$$

where k is an unknown numerical coefficient, which can be determined empirically (the result is $k \approx 0.2$).
Equations (9.42) and (9.43) yield

$$l(t) \propto \left(e|S|\frac{t}{\eta}\right)^{1/2}. \tag{9.48}$$

We have already mentioned the experimental results $R(t) \propto t^{0.75 \pm 0.05}$ and $l(t) \propto t^{0.4 \pm 0.1}$. Both results are in excellent agreement with the model. The agreement with the dependence on viscosity in equation (9.47) is not quite as good (the measured value is 0.6, as opposed to the calculated value of 0.75).
After the sandwiched film has dewetted, a dry contact of radius a forms between the elastic ball and the hydrophobic glass. H. Hertz was the first to calculate the size of the contact of an elastic ball pressed against a planar surface with a force F.[35] In the absence of any adhesion forces between the elastomer and glass, the radius a of the flat portion of the contact and the penetration depth δ of the ball can be determined by minimizing the energy of the squeezed ball:

$$G \propto E \left(\frac{\delta}{a}\right)^2 a^3 + F\delta \tag{9.49}$$

where $\delta \approx a^2/R$. Minimizing equation (9.49) with respect to a at fixed R gives

$$a^3 \approx \frac{FR}{E}. \tag{9.50}$$

Note that the surface area is the same whether the ball is in air or immersed in the liquid. The so-called JKR theory (Johnson, Kendall, and Roberts) extends Hertz's calculation to the case when the rubber adheres to the glass.[36] The theory takes Dupré's adhesion energy into account via the relation $W_0 = \gamma_{SL} + \gamma_{LR} - \gamma_{SR} = -S$. This energy depends on whether the ball sits in air or is immersed. It leads to the establishment of a contact even in the absence of an applied force. In other words, the contact surface area (or radius a) is not zero even when $F = 0$. If the force F is such that $F > |S|a^2$, Hertz's theory is applicable. That was the regime in which P. Martin did his work. Measuring a then yields the value of the normal component of the applied force F. This force, which is responsible for the drainage process studied in section 9.4.3.1 [equation (9.37)], must be overcome to lubricate a contact in a forced wetting regime. It is the object of the next section.

9.4.4   Wetting Transitions Under Shear: The Principle of Hydroplaning

Forced wetting at a solid/air interface is a classical phenomenon and was described in chapters 5 and 6. We have seen that when a plate (or fiber) is drawn out of a non-wetting liquid bath at increasing velocity U, the plate remains dry up to a threshold velocity V_c, and it is wetted by a film of thickness $e(U) \propto U^{2/3}$ as soon as $U > V_c$. The velocity V_c can be thought of as resulting from competition between forced wetting and dewetting, which leads to $V_c \approx \gamma \theta_E^3 / \eta$. The transition at V_c is discontinuous.

We focus here on forced wetting of a liquid at the interface between a hard solid and a soft one. This situation has begun to be understood only very recently, even though it controls the deposition of films by soft implements (paints, printing, etc.), as well as the loss of adhesive contacts (aquaplaning).

We now consider a situation in which we introduce a shearing stress on the bond between an elastomer and a glass surface. Suppose the glass surface slides at a constant velocity U along a fixed direction with respect to the stationary elastomer.[37] The contact between elastomer and glass starts out dry (following the spontaneous dewetting of the wedged-in liquid film).

Upon moving the elastomer ball at constant speed U on the substrate, A. Martin observed two distinct lubrication transitions at threshold velocities V_{c1} (~ 10 µm/s) and V_{c2} (~ 30 µm/s), as illustrated in Figure 9.31.[32]

1. For $U < V_{c1}$, the contact is dry and the system is in a dewetted state (Figure 9.31a).
2. For $V_{c1} < U < V_{c2}$, the contact becomes semi-lubricated, featuring dry zones together with liquid inclusions (Figure 9.31b).
3. For $U > V_{c2}$, the contact is fully lubricated (Figure 9.31c). The ball has then entirely lifted off the substrate and there no longer is contact. This is the aquaplaning regime.

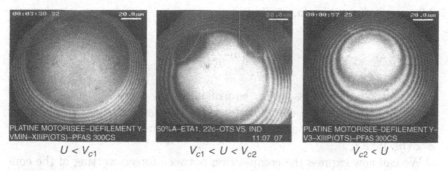

$$U < V_{c1} \qquad V_{c1} < U < V_{c2} \qquad V_{c2} < U$$

FIGURE 9.31. Three regimes of a sheared JKR contact: dry, semi-lubricated, and lubricated (courtesy A. Martin).

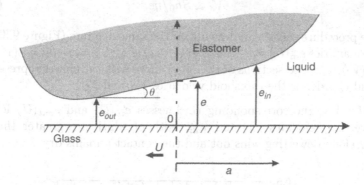

FIGURE 9.32. Profile of a ball in the lubricated regime: The tilted pad model.

The transitions at V_{c1} and V_{c2} can be interpreted as competition between invasion of the liquid brought on by shearing and spontaneous dewetting which opposes it.

Figure 9.32 illustrates the profile of the lubricated ball ($U > V_{c2}$). The sandwiched film, of average thickness e, forms a liquid wedge with a very small angle $\theta \approx e/a$ ($\sim 10^{-3}$ to 10^{-2} rad). The thicknesses of the wedge are e_1 and e_2 at the leading and trailing edge, respectively.

How Does the Thickness e Vary as a Function of the Velocity U?

We know the vertical hydrodynamic lift force exerted on the rubber viewed as a tilted planar pad:[25]

$$F_z \propto \eta \frac{U}{\theta^2} a\Delta \tag{9.51}$$

where $\theta \approx e/a$ and $\Delta = (e_2 - e_1)/e = cnst$. F_z offsets the penetration force F, which is related to the size of the flat portion of the contact via Hertz's formula [equation (9.50)]:

$$\eta \frac{U}{\theta^2} a\Delta \approx a^3 \frac{E}{R}. \tag{9.52}$$

After replacing θ by e/a, we find:

$$e \propto \left(\eta \frac{UR}{E} \right)^{1/2}. \tag{9.53}$$

The average thickness e (or, equivalently, the angle θ) varies as $U^{1/2}$ and is independent of the applied force F. This law turns out to be in fair agreement with the experimental results of Clain and Martin, who found $e \propto (\eta U)^{0.6} F^0$.

We can now express the competition between forced wetting of the contact and its dewetting. We have seen that the initial dewetting velocity is related to the thickness e of the film through the relation:

$$V_d \propto S h_0 / \eta e. \tag{9.54}$$

The procedure consists in drawing on the same diagram (Figure 9.33) the three quantities $e_{in}(U) = e + \theta a = e(1 + \alpha_1)$, $e_{out}(U) = e - \theta a = e(1 - \alpha_1)$, and $e(V_d)$. The intersections of the $e(V_d)$ curve with the curves representing e_{in} and e_{out} define the threshold velocities V_{c1} and V_{c2}.

- If $U < V_{c1}$, the corresponding thicknesses $e_{in}(U)$ and $e_{out}(U)$ lead to dewetting velocities V_{d1} and V_{d2}, both of which are greater than U. Therefore, dewetting wins out and the contact remains dry.

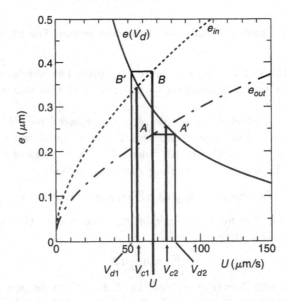

FIGURE 9.33. Geometrical determination of the velocities V_{c1} and V_{c2} separating the three lubrication regimes (dry, semi-lubricated, and fully lubricated). For a given velocity U, the two thicknesses $e_{out}(A)$ and $e_{in}(B)$ are determined and the corresponding dewetting velocities (at points A' and B') are compared to U. If $V_d > U$, the contact is dry; otherwise it is wet.

- If $V_{c1} < U < V_{c2}$, we get $V_{d1} > U$ and $V_{d2} < U$, which means that the trailing edge is the only one that has enough time to dewet (semi-lubricated regime).
- If $U > V_{c2}$, both V_{d1} and V_{d2} are smaller than U. In this case, dewetting cannot keep up and the contact remains fully wet, leading to a total loss of adhesion (aquaplaning).

Conclusion. A liquid film trapped at the interface between a solid and a rubber dewets by nucleation and growth of a dry zone. The final equilibrium state consists of a dry contact between solid and rubber. When such a dry contact is sheared, a first transition of the contact to a partial wetting condition occurs at a threshold velocity V_{c1} (semi-lubricated regime). Beyond a second, higher threshold velocity V_{c2}, wetting becomes total and contact between solid and rubber is lost altogether.

9.4.5 Role of Nucleators in Forced Wetting: Cerenkov Wake

We have seen that a car skids out of control upon braking suddenly on a wet road because a film of liquid invades the contact. Can adhesion be recovered with some suitable nucleator?

To answer the question, we can etch Vickers imprints of size b ($b \approx 10$ µm) on a glass surface. As we have seen, such imprints can initiate the dewetting of films with a thickness $e \leq e^* = \sqrt{h_0 b}$.[38]

When the imprint passes across the lubricated contact zone at velocity U, several regimes appear, as illustrated in Figure 9.34.

- At very high velocities ($U > U_{n2}$), the nucleator remains immersed, as the thickness $e(U)$ is greater than e^*.
- At intermediate velocities ($U_{n1} < U < U_{n2}$), a regime dubbed the "pear" condition sets in, in which the nucleator triggers dewetting in the middle of the contact. The dry zone adopts a triangle-shaped head with rounded edges at the tail—a pattern somewhat resembling a pear.

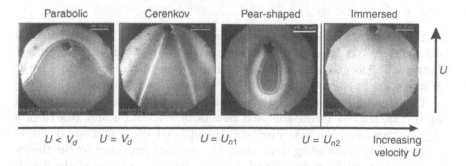

FIGURE 9.34. Wakes created by a defect engraved on glass as it traverses the lubricated contact at velocity $U(U > V_{c2})$ (courtesy A. Martin).

- At low velocities $(U < U_{n1})$, we are in the so-called Cerenkov regime. The nucleator triggers dewetting the moment it crosses the flat region of the contact. The wake forms a triangle of apex angle α_0 (Figure 9.34). It has been dubbed "Cerenkov wake" by analogy with the situation of a relativistic particle emitting a cone of light when it is moving at a velocity U greater than the velocity of light in the medium. Here, the nucleator triggers a dewetting wave. The wave progresses at a velocity V_d smaller than the nucleator velocity U, in accordance with the definition of the lubricated regime $(U > V_d$ everywhere). The wake is also reminiscent of that left behind by a supersonic aircraft. The wake's wedge angle, also known as the Mach angle, is given by the relation

$$\sin \alpha_0 = \frac{V_d}{U}. \tag{9.55}$$

Actually, the shape of the dry zone is not strictly triangular because of the ridge that draws liquid away from the drying zone and expands far from the nucleator (Figure 9.34), leading to a decreasing dewetting speed. Taking this effect into account, one can calculate the detailed profile of the dry zone.[32, 38]

The physical meaning of the velocities U_{n1} and U_{n2} is explained in Figure 9.33, where we have plotted the quantities e_{in} and e_{out} as functions of U. To a defect of size b we can associate a thickness $e^* = \sqrt{h_0 b}$. The velocity U_{n1} is defined by the equality $e^* = e_{in}(U_{n1})$, and U_{n2} by the equality $e^* = e_{out}(U_{n2})$. For $U > U_{n2}$, the thickness of the film in its flat zone is everywhere greater than e^*, and the defect remains submerged. For $U_{n1} < U < U_{n2}$, the thickness e^* is less than e_{in}, and dewetting can be initiated at the center of he contact. For $U < U_{n1}$, e is everywhere less than e^* and dewetting begins as soon as the defects enters the flat zone. The result is then a Cerenkov wake.

With these considerations in mind, we can show that large nucleators have the ability to prevent a car from skidding on a wet road by restoring dry zones.

9.4.6 Conclusion

In this chapter, we have seen how a sandwiched film can promote adhesion on a surface. We have examined the establishment of an adhesive contact between a deformable elastomer lens and a rigid planar solid—whether it is in motion or at rest—immersed in a non-wetting liquid.

The liquid film (fluorinated oils, water-glycerol mixtures, etc.) trapped at the interface is metastable and dewets by nucleation and growth of a dry contact zone. We have discussed the optimal shape and size of nucleators capable of initiating dewetting. The relation predicted by the theory between the critical radius for nucleation and the thickness of the film is consistent with experimental results.

At equilibrium, following the dewetting of a wedged-in film, the contact between the elastomer lens and the hydrophobic lens is dry. We have studied the forced wetting of the contact induced by shearing, when the glass slide is moved at a constant velocity U. There are two wetting transitions during which the contact is invaded, first partially and then totally, by the liquid. These transitions have been interpreted in terms of a competition between forced wetting induced by shearing and dewetting. At high speed, the contact is wetted by a film of thickness $e(U)$ (aquaplaning) calculated with the help of a simple hydrodynamic model.

Finally, we have studied the role of nucleators on films in forced wetting situations. Such nucleators can be created by etching defects of size b on glass, which pass through the lubricated contact at constant velocity U. The defects remain submerged at high velocities. However, for more modest velocity values, the defects leave in their wake triangular dry zones, which progress at the dewetting velocity V_d. We have shown that just a few "large" nucleators have the ability to prevent aquaplaning, that is to say, the total loss of adhesive contact.

These examples of forced wetting and dewetting, which we studied in ideal cases, concern more than just the art of driving safely on wet roads. They are relevant to all adhesion phenomena on wet substrates, including boat glues and cellular adhesion. Figure 9.35 shows a mushroom spore adhering on a wet and highly hydrophobic rice leaf.

Cellular adhesion controls cellular development [e.g., pyriculariosis, also known as rice blast (Figure 9.35), embryos, tumors, and so forth] as well as intercellular communications (exchange of signals, of proteins, etc.). This

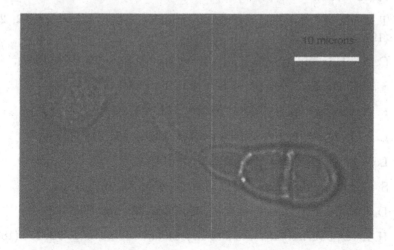

FIGURE 9.35. Spore of *magnaporthe grisea* adhering strongly to a highly hydrophobic rice leaf. (From F. Brochard, A. Buguin, P. Martin, and O. Sandre, in *Journal of Physics: Condensed Matter, 12A*, p. 239 (2000). IOP Publishing. Reproduced by permission.)

field is currently in rapid development. It is therefore important to understand not only the specific interactions responsible for adhesion, but also the dynamics of contact formation.

References

[1] R. E. Johnson and R. H. Dettre, in Contact Angle, Wettability, and Adhesion, *Advances in Chemistry Series* **43**, 112 (1964).

[2] R. N. Wenzel, *Ind. Eng. Chem.* **28**, 988 (1936).

[3] A. B. D. Cassie and S. Baxter, *Trans. Faraday Soc.* **40**, 546 (1944).

[4] R. Lipowsky, *Current Opinion in Colloid and Interface Science* **6**, 40 (2001).

[5] J. Bico, C. Tordeux, and D. Quéré, *Europhys. Lett.* **55**, 214 (2001); J. Bico, C. Marzolin, and D. Quéré, *Europhys. Lett.* **47**, 220 (1999); J. Bico, *Doctoral thesis, University of Paris* (2000).

[6] A. Lafuma and D. Quéré, *Nature Materials* **2**, 457 (2003); J. Bico, U. Thiele, and D. Quéré, *Colloids and Surfaces A*, 206, 41 (2002).

[7] J. F. Oliver, C. Huh, and S. G. Mason, *J. Colloid Interface Sci.* **59**, 568 (1977).

[8] S. Shibuichi, T. Onda, N. Satoh, and K. Tsujii, *J. Phys. Chem.* **100**, 19512 (1996).

[9] T. Onda, S. Shibuichi, N. Satoh, and K. Tsujii, *Langmuir* **12**, 2125 (1996).

[10] C. Neinhuis and W. Barthlott, *Annals of Botany* **79**, 667 (1997).

[11] S. Herminghaus, *Europhys. Lett.* **52**, 165 (2000).

[12] P. Aussillous and D. Quéré, *Nature* **411**, 924 (2001).

[13] L. Mahadevan and Y. Pomeau, *Phys. Fluids* **11**, 2449 (1999).

[14] Lord Rayleigh, *Phil. Mag.* **28**, 161 (1914).

[15] S. Chandrasekhar, *Proc. Roy. Soc. London* A**286**, 1 (1965).

[16] D. Richard and D. Quéré, *Europhys. Lett.* **50**, 769 (2000).

[17] H. Hertz, *Journal für reine und angewandte Mathematik* **92**, 156 (1881).

[18] D. Richard, C. Clanet, and D. Quéré, *Nature* **417**, 811 (2002).

[19] Lord Rayleigh, *Proc. Roy. Soc. London* **29**, 71 (1879).

[20] A. Marmur, *J. Colloid Interface Sci.* **122**, 209 (1988).

[21] E. Raphaël and P. G. de Gennes, *C. R. Acad. Sci. (Paris)* **327 IIB**, 685 (1999).

[22] L. Bacri and F. Brochard-Wyart, *Europhys. J. E* **2**, 87 (2000).

[23] A. Aradian, E. Raphaël, and P. G. de Gennes, *Europhys. J. E* **2**, 367 (2000).

[24] A. D. Roberts and D. Tabor, *Proc. Roy. Soc. London A* **325**, 323 (1971).

[25] J.-M. Georges, *Frottement, usure et lubrification* (Friction, Wear-out and Lubrication) (Paris: Eyrolles, 2000); Frank Philip Bowden and D. Tabor, *Friction and Lubrication of Solids* (Oxford, U.K., New York: Oxford University Press, 2000).

[26] L. D. Landau, *Theory of Elasticity* (Oxford, U.K., New York: Pergamon Press, 1982).

[27] I. N. Sneddon, *Proc. Roy. Soc. London A* **187**, 229 (1946).

[28] P. Martin, P. Silberzan, and F. Brochard-Wyart, *Langmuir* **18**, 4910 (1997).

[29] P. Martin and F. Brochard-Wyart, *Phys. Rev. Lett.* **80**, 3296 (1998).

[30] P. Martin, *Doctoral thesis, University of Paris* (1997).

[31] A. D. Roberts, *J. Phys. D* **4**, 3 (1971).

[32] A. Martin, *Doctoral thesis, University of Paris* (2000).

[33] A. Martin, A. Buguin, and F. Brochard-Wyart, *Langmuir* **17**, 6553 (2001).

[34] F. Brochard-Wyart and P. G. de Gennes, *J. Phys. Condens. Matter* **6A**, 9 (1994).

[35] H. Hertz, *Miscellaneous Papers* (London: McMillan and Co., 1896).

[36] K. L. Johnson, K. Kendall, and A. D. Roberts, *Proc. Roy. Soc. London* **A324**, 301 (1971).

[37] A. Martin, J. Clain, A. Buguin, and F. Brochard-Wyart, *Phys. Rev. E* **65**, 031605, (2002).

[38] A. Martin, A. Buguin, and F. Brochard-Wyart, *Europhys. Lett.* **57**, 71 (2002).

10
Transport Phenomena

10.1 Chemical Gradients

If the chemical composition of an interface varies from point to point, so does the surface tension creating forces that drag the liquid in a direction parallel to the substrate.

10.1.1 Experiments With Vapors

There exists a rich folklore revolving around the movements of liquids induced by an inhomogeneous chemical environment.

Example 1: Wine tears. In a glass of wine (mixture of water and alcohol), we can readily observe a film rising up along the wall (Figure 10.1). The film terminates at its topmost part in a ridge, which breaks up into "tears" falling back down under their own weight. The phenomenon is intriguing: Why does a film rise in the first place?

The answer has to do with the fact that alcohol evaporates faster in the film (which has a higher surface-to-volume ratio). As it turns out, alcohol lowers the surface tension of water. As a result, the higher part of the film, which is alcohol deprived, has a higher surface tension. It is this excess of surface tension that draws the film upwards.*

---

*There is also another effect. As the alcohol evaporates, the film cools down, resulting in a slightly higher surface tension due to a thermal effect, which reinforces the chemical effect.

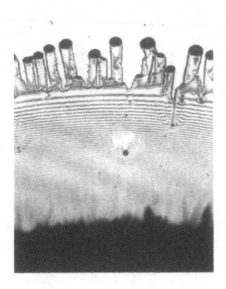

FIGURE 10.1. A manifestation of the Marangoni effect: Wine tears. (From A. M. Cabazat and J. B. Fournier, in *Bulletin SFP, 84,* p. 22 (1992). Reproduced by permission.)

Example 2: Drops chasing each other. (Figure 10.2). Imagine that we place side by side a drop of PDMS and a drop of solvent (for instance transdecaline), as depicted in Figure 10.2. The solvent is volatile and raises the surface tension of PDMS upon mixing with it. The surprising outcome is that the drop of PDMS starts moving toward the drop of solvent. Maxwell observed this type of phenomenon way back in 1878. The principle is actually quite simple. The decaline evaporates and lands primarily on the near edge of the PDMS drop, at point A in Figure 10.2. As a result, the surface tension at point A on the near edge is greater than at point B on the far edge, and the entire drop is set in motion toward the decaline.

This effect has been studied more systematically in the geometry illustrated in Figure 10.3, where the solvent is arranged along a ring surrounding the oil drop, with the entire setup placed in an isolated enclosure.[1] In this case, because of the circular symmetry, the drop expands equally in all directions.

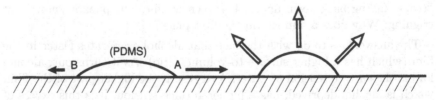

FIGURE 10.2. Interactions between two drops. The drop of PDMS is at left, that of transdecaline is at right. The solvent evaporates (large arrows) and lands primarily on the near side of the PDMS, causing it to spread at different velocities all along its perimeter (small arrows).

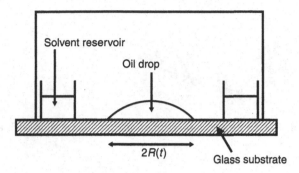

FIGURE 10.3. Principle of experiments with volatile solvents.

10.1.2 Transport Toward Wettable Regions

While fabricating special surfaces by silanizing only half of a glass slide, C. Casagrande noticed that a drop of water placed at the boundary between the two regions on the hydrophobic side rushes over to the hydrophilic side. Raphaël and Ondarçuhu have studied this phenomenon quantitatively.[2] It is confined to the transition region and the shape of drops is complicated.

To generate spontaneous movements over larger scales, one can use a glass slide with a wettability gradient. Such slides have been developed in biotechnology to separate proteins according to their degree of hydrophobicity. On a slide free of hysteresis, a drop driven by the gradient can move toward the most wettable region. Unfortunately, a drop of water placed on the hydrophobic side often refuses to move because the chemical gradient is too weak to overcome hysteresis.

Chaudhury and Whitesides succeeded in making hysteresis-free slides with a wettability gradient by exposing a silicon wafer to the diffusion front of decyltrichlorosilane vapor, of formula $Cl_3–Si–(CH_2)_9–CH_3$.[3] With this technique, they managed to create on the slide a concentration gradient and, accordingly, a hydrophobicity gradient giving rise to a contact angle with water varying from 10° to 25° over a distance of a centimeter. There is a weak hysteresis ($\theta_A - \theta_R < 10°$).

When the wafer is tilted with the hydrophobic part at the bottom, a small drop of water is able to climb up the incline at speeds of the order of a mm/s (Figure 10.4).

The drop moves because the contact angle θ_{E_A} at the trailing edge (A) is greater than the angle θ_{E_B} at the leading edge (B). Therefore, the Laplace pressure is greater at the trailing edge than at the leading one, setting the drop in motion (Figure 10.5).

We shall now examine more quantitatively the motion of drops and puddles on horizontal surfaces bearing the imprint of a wettability gradient in the x-direction. Both $\gamma_{SO}(x)$ and $\gamma_{SL}(x)$ depend on the position x, and therefore, so does the spreading parameter $S(x)$. We consider the case when $S' = dS/dx$ is small.

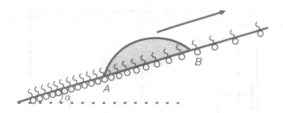

FIGURE 10.4. PDMS drop climbing up an inclined glass slide with a wettability gradient.

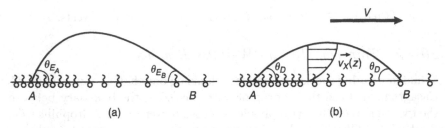

FIGURE 10.5. Drop on a surface with a wettability gradient. When the drop is at rest (a), its trailing edge, where the curvature is the highest, experiences an overpressure. The asymmetry sets the drop in motion (b).

General Description of the Motion

The motion is determined by balancing the capillary driving force against the viscous resistance force of the drop in motion.[4] We start by studying the motion of *ribbons* of width l (along the x-axis) and extending in a direction parallel to the y-axis. This may occur in either a capillary regime $(l < \kappa^{-1})$ or a gravity regime $(l > \kappa^{-1})$ (Figure 10.6). We shall subsequently extend the analysis from second ribbons to third drops.

Driving Force F_D. If we displace the ribbon by a distance dx, the surface energy U varies by dU:

$$dU = [(\gamma_{SL} - \gamma_{SO})_B - (\gamma_{SL} - \gamma_{SO})_A]\, dx. \tag{10.1}$$

This generates a driving force F_D (per unit length) given by

$$F_D = -\frac{dU}{dx} = -(\gamma_{SL} - \gamma_{SO})_B + (\gamma_{SL} - \gamma_{SO})_A. \tag{10.2}$$

Viscous Resistance Force F_V. The friction force F_V (also per unit length) is the integral of the viscous stress σ_{xz} within the liquid/solid interface $(z = 0)$:

$$F_V = \int_A^B \sigma_{xz}(0)\, dx. \tag{10.3}$$

In the limit of small angles, when we can use the lubrication approximation, the velocity profile $v_x(z)$ of the flow is of the Poiseuille type because

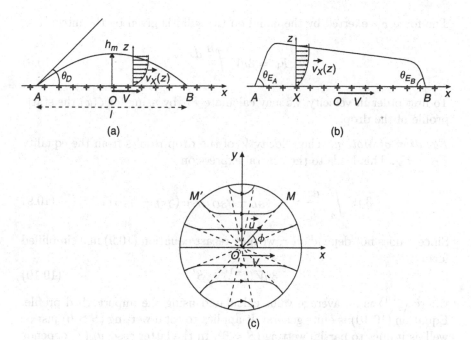

FIGURE 10.6. Drops in motion on a chemically inhomogeneous surface. (a) Ribbon with a circular cross section; (b) "mesa" flattened by gravity; (c) circular drop with spherical surface.

the pressure $P(x)$ inside the drop depends on x but not on the vertical coordinate z. Therefore,

$$\eta v_x(z) = \frac{1}{2}\frac{dP}{dx}(x^2 - 2z\zeta) \qquad (10.4)$$

which satisfies the boundary conditions $v_x(0) = 0$ and $(dv_x/dz)_{z=\zeta} = 0$ at the liquid air interface located at height $\zeta(x)$. At this stage, there is no need to write $P(x)$ explicitly. Instead, we will express the results in the terms of the translation velocity V of the drop. The flow rate of matter is given by

$$V\zeta = \int_0^\zeta v_x(z)\, dz. \qquad (10.5)$$

With the help of equation (10.5), we can eliminate the derivative dP/dx appearing in equation (10.4). The result is

$$v_x(z) = \frac{3V}{2\zeta^2}(-z^2 + 2z\zeta). \qquad (10.6)$$

The viscous stress at the liquid/solid interface is therefore

$$\sigma_{xz}(0) = 3\eta\frac{V}{\zeta}. \qquad (10.7)$$

The force F_V exerted by the liquid on the solid is given by the integral

$$F_V = 3\eta V \int_A^B \frac{dx}{\zeta}. \tag{10.8}$$

To first order in velocity, we may calculate F_V by using for $\zeta(x)$ the static profile of the drop.

Equation of Motion. The velocity V of the drop results from the equality $F_D = F_V$. This leads to the general expression

$$3\eta V \int_A^B \frac{dx}{\zeta} = -(\gamma_{SL} - \gamma_{SO})_B + (\gamma_{SL} - \gamma_{SO})_A. \tag{10.9}$$

Since γ does not depend on x, we may recast equation (10.9) in a simplified form

$$3\eta V \langle \zeta^{-1} \rangle = S' \tag{10.10}$$

where $\langle \zeta^{-1} \rangle$ is an average value calculated using the unperturbed profile. Equation (10.10) is quite general. It applies to total wetting $(S > 0)$ just as well as it does to partial wetting $(S < 0)$. In the latter case, $\theta_E(x)$ depends on the spatial coordinate x. In accordance with Young's equation, we have

$$\sin^2 \left[\frac{\theta_E(x)}{2} \right] = -\frac{S(x)}{2\gamma}. \tag{10.11}$$

We will use equation (10.10) to describe the movement induced by a gradient $S' = dS/dx$ for several geometries, namely, a one-dimensional liquid line, a ribbon flattened by gravity, and a spherical drop (Figure 10.6).

Ribbon with Circular Cross Section (Figure 10.6a). Here, the friction force is dominated by the two opposite liquid wedges:

$$F_V = 3\eta V \int_A^B \frac{dx}{\zeta} = 6\eta V \int_A^B \frac{dx}{(x - x_A)\theta_{E_0}} = 6\eta \frac{V}{\theta_{E_0}} \ln \tag{10.12}$$

where $\ln \equiv \ln(l/a)$ describes the divergence of the viscous dissipation in the wedge (see chapter 6), a is a molecular size, and θ_{E_0} is an average angle. Equating this force to the driving force $F_d = lS'$, we find

$$V = \frac{l\theta_{E_0}}{6\eta \ln} S' = \frac{2h_m S'}{3\eta \ln} \tag{10.13}$$

where h_m is the height at the center of the ribbon.

Mesa-Shaped Flat Ribbon (Figure 10.6b). For flat structures, the dissipation is dominated not by the edges but, rather, by the inner part of the drop. The corresponding force F_V may be calculated [see equation (2.10)] by assuming a constant thickness $h_m = 2\kappa^{-1}\sin(\theta_{E_0}/2)$:

$$F_V = 3\eta V \int_A^B \frac{dx}{\zeta} = 3\eta V \frac{l}{h_m}. \tag{10.14}$$

This expression is valid as soon as $l > \kappa^{-1}\ln$. The expression for the velocity then becomes

$$V = \frac{h_m S'}{3\eta}. \tag{10.15}$$

Upon comparing equations (10.13) and (10.15), we observe that the velocity of large drops is enhanced by a factor $1/2\ln \approx 7$; in other words, large drops move significantly faster than small ones.

Figure 10.6b shows the mesa moving at a velocity V. Because the wedge dissipation is negligible, the contact angles at points A and B can be replaced by θ_{E_A} and θ_{E_B}, respectively. The free surface is no longer perfectly horizontal. This creates a gradient of the pressure $P(x)$ driving the drop toward more wettable regions. Ondarçuhu has actually observed the corresponding slope.

Spherical Drops (Figure 10.6c). This case is more complicated. A rough estimate of the velocity is[4]

$$V = \frac{2h_m S}{3\eta \ln} \tag{10.16}$$

where h_m is the thickness at the center of the spherical cap.

| **Problem:** If $\cos\theta$ goes from 0 to 1 over a distance of 5 cm, what is the velocity of a water drop characterized by $l = 1$ mm and an angle $\theta_0 = 60°$?

Van der Waals Pancakes ($S > 0$). For these flat structures, the dissipation is again governed by the inner portion of the drop. The thickness e of the drop is related to S by $e = a\sqrt{3\gamma/2S}$. The velocity of these miniature puddles is the same as that of gravity puddles after replacing h_m by e_0, which is the thickness calculated by making S equal to S_0 at the center:

$$V = \frac{e_0 S'}{3\eta}. \tag{10.17}$$

Conclusion. Drops always move toward regions with larger values of S in order to lower their surface energy. They do so with a velocity V given by

$$V \propto \frac{S'}{\eta} h_m. \tag{10.18}$$

This holds for any geometry and any size in the viscous regime.

Note: Role of Hysteresis

In the case of non-perfect surfaces, hysteresis creates a force $\delta = \gamma(\cos\theta_R - \cos\theta_A)$, which opposes any motion. Movement can begin only at a threshold gradient S' since the driving force can then be written as:

$$F = lS' - \delta. \tag{10.19}$$

The velocity of a drop on a "dirty" surface then becomes

$$V \propto \frac{h_m}{\eta}\left(S' - \frac{\delta}{l}\right). \tag{10.20}$$

10.2 Thermal Gradients

10.2.1 Drops Favoring the Cold

Brzoska has studied the movement of PDMS drops placed on silanized silicon wafers (OTS) subjected to a horizontal temperature gradient.[5] The crucial element of the experiment was the preparation of the substrate, since high hysteresis would prevent drops from moving.

In the following experiments, the advancing and receding angles were 13° and 11°, respectively $[\delta = \gamma(\cos\theta_R - \cos\theta_A) \approx 10^{-2}]$. Under such conditions, drops can be made to move even in weak thermal gradients. This could never be done with chemical gradients because of their greater hysteresis.

The thermal gradient modifies all three surface tensions $\gamma_{SO}[T(x)]$, $\gamma_{SL}[T(x)]$, and $\gamma[T(x)]$. Therefore, the spreading parameter S will depend on x, as was the case in the previous example of surfaces exhibiting a built-in wettability gradient. However, there is a new effect, namely, the inhomogeneity of the surface tension $\gamma(x)$, which will bring about a stress at the free surface of the liquid. For PDMS, we have $\frac{1}{\gamma}\frac{d\gamma}{dT} = -2.5 \cdot 10^{-3}$ K^{-1}. A flow will then be set up toward the colder regions because that is where γ is the greatest. We are thus faced with two motion-inducing phenomena:

1. The wettability gradient, which tends to drive drops toward regions of higher S;
2. the "Marangoni effect," which tends to drive drops in the opposite direction, toward the colder regions.

It is impossible to predict a priori the direction in which the drops will end up moving.

The experimental setup used to study the problem is depicted in Figure 10.7. The silicon wafers are rectangular to obtain temperature gradients ∇T that are uniform, ranging from 0.1 to 1°/mm.

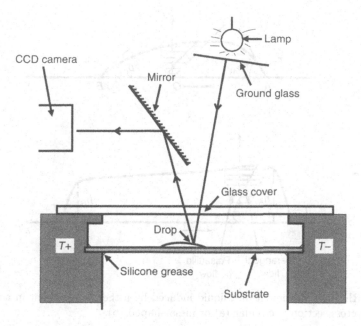

FIGURE 10.7. Experimental setup for studying the motion of drops under the influence of a thermal gradient (courtesy F. Rondelez).

Experiments lead to the following observations:

- Drops move toward the colder region if their radius is greater than a critical value R_c that depends on the hysteresis δ of the contact angle and is inversely proportional to ∇T.
- Beyond the critical value R, the velocity V increases linearly with R and ∇T.
- For puddles flattened by gravity, the velocity of drops saturates at a value that is independent of their size. Their periphery is no longer circular but, rather, exhibits two straight segments aligned in the direction of the thermal gradient.

We describe next the movement of such drops on ideal surfaces.[4] (We will not consider the influence of hysteresis until later.) Here again, we express the equilibrium between the driving force and the viscous dissipation force. For the sake of simplicity, we restrict ourselves to the case of a one-dimensional ribbon of length l, whose cross section is either circular ($l < \kappa^{-1}$) or mesa-shaped ($l > \kappa^{-1}$), as illustrated in Figure 10.8.

The driving force is unchanged since it is defined by a displacement dx of the drop and involves only γ_{SL} and γ_{SO}:

$$F_D = (\gamma_{SO} - \gamma_{SL})_B - (\gamma_{SO} - \gamma_{SL})_A = l \left(\frac{dS}{dT} + \frac{d\gamma}{dT} \right) \frac{dT}{dx}. \quad (10.21)$$

$S(T)$ is determined experimentally from measurements of $\theta_E(T)$. The

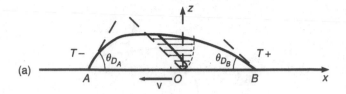

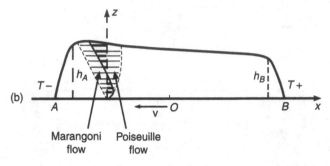

FIGURE 10.8. Movement of a liquid induced by a thermal gradient in a ribbon whose cross section is circular (a) or mesa-shaped (b).

friction force, on the other hand, is modified by the Marangoni effect, which induces flows within the drop toward the cold region. Therefore, the net liquid flow is a superposition of a Poiseuille flow created by the pressure gradient dP/dx and of a simple shear stress due to $d\gamma/dx$ exerted on the free surface of the liquid.

The boundary condition for the velocity at the free surface is now $\eta \partial v_x / \partial x = \partial \gamma / \partial x$. The velocity profile depicted in Figure 10.8 is therefore given by

$$\eta v_x(x, z) = \frac{1}{2} \frac{dP}{dx} (z^2 - 2z\zeta) + \frac{\partial \gamma}{\partial x} z. \tag{10.22}$$

The viscous stress at the solid/liquid interface is

$$\sigma_{xz}(0) = -\frac{dP}{dx} \zeta + \frac{\partial \gamma}{\partial x} \tag{10.23}$$

where $\frac{d\gamma}{dx} = \frac{\partial \gamma}{\partial T} \frac{\partial T}{\partial x}$. Integrating $v_x(x, z)$ yields the flow $J(x)$

$$\eta J(x) = -\frac{1}{3} \frac{dP}{dx} \zeta^3 + \frac{\partial \gamma}{\partial x} \frac{\zeta^2}{2}. \tag{10.24}$$

Writing that the liquid is stationary in the frame of reference of the substrate moving at a velocity $-V$ leads to $J = \zeta V$. We can eliminate dP/dx in equation (10.24), which gives

$$\sigma_{xz}(0) = 3\eta \frac{V}{\zeta} - \frac{1}{2} \frac{\partial \gamma}{\partial x}. \tag{10.25}$$

The force of friction acting on the substrate is therefore

$$F_V = \int_A^B 3\eta \frac{V}{\zeta}\, dx - \frac{1}{2}(\gamma_A - \gamma_B). \qquad (10.26)$$

The Marangoni contribution is the second term in equation (10.26). It leads to a displacement of drops toward regions of high surface tension γ, in other words, toward the cold side.

The equation of motion is still given by

$$F_V = F_D. \qquad (10.27)$$

For circular or mesa-shaped ribbons, we find with the help of equations (10.21) and (10.26):

$$V = V^* h_m \left[\frac{1}{\gamma} \frac{d(\gamma_{SO} - \gamma_{SL})}{dT} + \frac{1}{2\gamma} \frac{d\gamma}{dT} \right] \frac{dT}{dx} \qquad (10.28)$$

where $V^* \approx \gamma/\eta$. This last expression shows that the velocity of drops increases with their radius R since $h_m \approx R\theta_E/2$ and saturates for puddles (when $h_m = cnst$).

If there is no contribution from the solid $[F_D = 0, d(\gamma_{SO} - \gamma_{SL})/dT = 0]$, the drops will move toward higher values of γ, which means toward the cold, where S is smaller. Therefore, the Marangoni effect drives drops in an unexpected direction. The direction of the motion depends on both $S(T)$ and $\gamma(T)$.

To include hysteresis, the equilibrium of forces must be written in the form

$$F_V = F_D - \gamma\delta \qquad (10.29)$$

where $\delta \equiv \cos\theta_R - \cos\theta_A$. The critical radius R_c is obtained by letting $V = 0$, which leads to

$$R_c = \frac{\gamma\delta}{\left(\dfrac{dS}{dT} + \dfrac{3}{2} \dfrac{d\gamma}{dT} \right) \dfrac{dT}{dx}}. \qquad (10.30)$$

This expression is in agreement with Brzoska's observations.

10.2.2 Finger Formation

Brzoska repeated the same experiment with a non-silanized silicon wafer, in which case the PDMS spreads completely.[6] He deposited a ribbon and let it spread at room temperature. As the ribbon reached a width of about 1 cm and a uniform thickness of 100 μm, he subjected the sample to a thermal gradient perpendicular to the ribbon. He saw the line of contact on the cold side develop undulations with a wavelength λ. The amplitude of the undulations increased exponentially with time and fingers formed, with a drop at each extremity, somewhat like a fingernail (Figure 10.9).

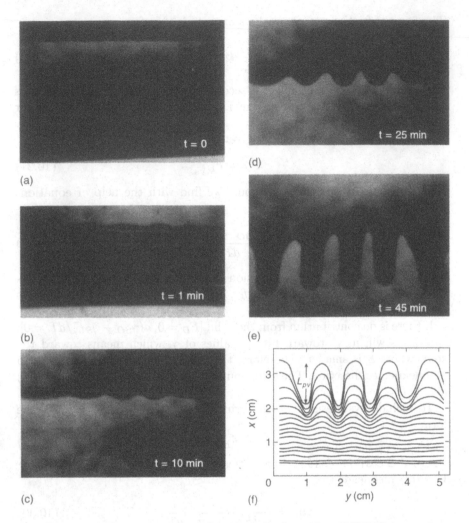

FIGURE 10.9. Time evolution of the edge of a fine ridge of PDMS exposed to cold. The ridge had a thickness of 16 μm and width of 2 cm and was subjected to a thermal gradient of 6.7°C/cm. The snapshots were taken at $t = 0$ (a); 1 min (b); 10 min (c); 25 min (d); and 45 min (e). Panel (f) shows the superposition of the profiles at different times. (From J. B. Brzoska, F. Brochard, and F. Rondelez, in *Europhysics Letters, 19* (2), p. 97 (1992). © EDP Sciences. Reproduced by permission.)

Suppose that the ribbon, in the process of spreading very slowly $[V_S \propto V^*(\kappa h_0)^3$, with $V^* = \gamma/\eta]$, is immersed in a thermal gradient. Its free surface is the seat of a stress $d\gamma/dx$, creating a shear flow of velocity $\eta v(z) = z \, d\gamma/dx$ and a material flow $J = \int v(z) \, dz = h_0 V_0$, where V_0 is given by

$$V_0 = \frac{1}{2} V^* \frac{1}{\gamma} \frac{d\gamma}{dx} h_0. \qquad (10.31)$$

If $V_0 > V_S$, a ridge forms at the line of contact and grows until the dynamical angle θ_D becomes such that the advancing velocity $V_S = V^*\theta_D^3$ equals V_0. At that point, the profile of the ridge becomes stationary in the frame of reference of the line of contact. It is then governed by equation (10.24) in which $J = eV_0$ and $P = -\gamma\zeta''$:

$$V_0 = \frac{1}{2}V^*\frac{1}{\gamma}\frac{d\gamma}{dT}\frac{dT}{dx}\zeta + \frac{1}{3}V^*\zeta^2\frac{d^3\zeta}{dx^3}. \tag{10.32}$$

The net flow is the sum of the Marangoni flow (first term) and the Poiseuille flow created by the gradient of the Laplace pressure due to the curvature of the ridge (second term). The flat region ($d^3\zeta/dx^3 = 0$) remains at constant thickness h_0, while the curved region (i.e., the ridge) is dominated by the Laplace pressure and has a characteristic size L_0, which is determined by comparing V_0 to the last term of equation (10.32):

$$L_0 = h_0\left(\frac{V^*}{3V_0}\right)^{1/3}. \tag{10.33}$$

A ridge is inherently unstable and breaks up into droplets. The thickness of the ridge governs the wavelength λ of the instability of the line (Figure 10.9):

$$\lambda = kL_0 \tag{10.34}$$

where k is a numerical constant.

Likewise, we can define a characteristic time τ_0 by way of a dimensional analysis:

$$\tau_0 = \frac{L_0}{V_0} \propto h_0^{-1/3}. \tag{10.35}$$

The time constant of the instability is then

$$\tau = m\tau_0 \propto h_0^{-1/3} \tag{10.36}$$

where m is a numerical constant.

Figure 10.10 shows the variation of λ and τ determined experimentally as functions of the initial thickness h_0. Experiments lead to $k = 22$ and $m = 6$, in agreement with detailed calculations.[7]

When the ridge begins to fragment into droplets, we are faced again with the problem of the motion of a drop in a thermal gradient, as discussed in the preceding section. Since the velocity of drops immersed in a temperature gradient ∇T is proportional to the height h_0, the thicker regions move faster and fingers appear, as shown in Figure (10.9).

The formation of fingers occurs only if the liquid is subjected to an external thermal gradient. It was first observed by A. M. Cazabat et al., who studied the rise of a liquid film up a *vertical* wall subjected to a thermal gradient.[8] To be sure, the vertical configuration is more complex because of gravity. In this case, the thickness h_0 results from competition between the Marangoni effects, which drives the flow upwards (toward the coldest region), and gravity, which drives it toward the bottom. In contrast to the

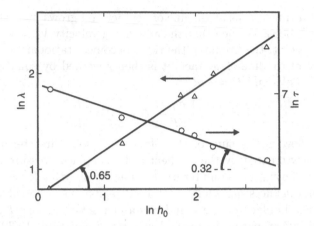

FIGURE 10.10. Dependence of the wavelength λ and the characteristic time τ of the periodic instability of the liquid front as functions of the initial thickness h_0 of the film.

horizontal configuration, where we start with a given film at time $t = 0$, in the present case a film forms and begins to undulate all at once. Since the two phenomena are coupled, it becomes very difficult to define the characteristic time of the instability, although the spatial period (which is much smaller here, being of the order of a millimeter) is still related to the thickness of the ridge as expressed in equation (10.34). The digitation of the line belongs in a family of instabilities of liquid films under conditions of forced spreading induced by an external force, such as a liquid ribbon

FIGURE 10.11. Festoon instability. Snapshot of a drop of low-volatility silicone oil (2-μ volume, 410-g molecular mass). Optical interference can be seen at the surface of the drop. A liquid ridge forms around the periphery of the drop and the line has a spatially modulated pattern. (From C. Redon, F. Brochard, and F. Rondelez, in *Journal de Physique II France*, 2, 1671 (1992). © EDP Sciences. Reproduced by permission.)

coming down an inclined plane or a liquid drop spreading by centrifugation (spin coating).[9,10] In both cases, a ridge forms near the line of contact, which ultimately leads to the formation of fingers.

If, on the other hand, the thermal gradient is not applied externally but is created by evaporation, the drops become festooned around their edges, as shown in Figure 10.11.[11] The drops do not, however, grow fingers because no external force is applied to them.

10.3   Reactive Wetting

10.3.1   Examples

Up to this point, we have dealt exclusively with *passive* liquids—liquids that do not alter the surface of the substrate. We now consider the case of *active* liquids, of which there are two broad categories:

1. those that etch the surface (such as a solution of nitric acid attacking a copper plate),
2. those able to deposit some substance on the surface.

We shall restrict our attention to the second type. Examples include the following:

- If we have a solution of surfactants exposed to a hydrophilic surface, it is possible for the surfactant to attach itself by its head onto the surface, transforming it into a brush of aliphatic chains, far less wettable than the original surface. Such a possibility has been known for some time, not only in solution but in pure liquids as well. A liquid of this type is said to be *autophobic*.[12]
- A chemical reagent in a liquid can attach itself to the surface and alter the wettability. For instance, a fatty acid in solution can bond with OH groups on the surface of silica (or glass) and create a hydrophobic monolayer. This case has been studied notably by Bain and coworkers.[13] The effect can be enhanced by using acids containing perfluorinated chains, which generate surfaces with exceedingly poor wettability toward water.

Another example studied by Ondarçuhu and Dominguez dos Santos involves a cholorosilane in an organic solution.[14] Here too a reaction occurs with OH groups on the surface. In its simplest form, the reaction is

$$-OH + Cl\text{-}Si\text{-}O\text{-}R \rightarrow -O\text{-}Si\text{-}O\text{-}R$$

Two situations can come up:

- If the reaction enhances wettability, a deposited drop spreads more readily as time progresses, but nothing particularly spectacular happens.

- If the surface becomes less wettable, the drop begins to "hate" the spot where it happens to rest on the substrate and wants to escape. We end up with what has come to be known as *running drops*, which move straight ahead at a constant velocity V. This is the situation on which we will focus our attention. However, before tackling the case of running drops, we start with the simpler case of a column of reactive liquid in a capillary.

We note in passing that similar phenomena occur with liquid (as opposed to solid) substrates.[15] However, a mathematical description of a liquid/liquid system is much more complicated from a hydrodynamics standpoint because of the flow induced in the underlying liquid.

10.3.2   Liquid Column in a Capillary

Consider a horizontal capillary tube containing a liquid column of length L very much larger than the tube's diameter d. Assume that the column moves toward the right at velocity V (Figure 10.12). At the leading edge (point B), no reaction has taken place yet. The equilibrium angle is Young's angle, which we designate θ_0. When in motion, the leading edge will exhibit a dynamical angle $\theta_B > \theta_0$ given by the usual relation:

$$F_B = \gamma(\cos\theta_B - \cos\theta_0) = \zeta_V V \tag{10.37}$$

where ζ_V is a friction coefficient discussed in Chapter 6. Let us now focus on the trailing edge (point A). Here, the equilibrium angle θ_{Ae} is different from θ_0 because the chemical reaction has altered $\gamma_{SL} - \gamma_{SO}$.

We may postulate that the change in the force is proportional to the fraction Φ of the surface that has reacted. This fraction is Φ_A at point A, and we may assume

$$\gamma_{SL} - \gamma_{SO} = (\gamma_{SL} - \gamma_{SO})_{\Phi=0} + \Phi_A \gamma_1 \tag{10.38}$$

where γ_1 is a coefficient characterizing the reduction in wettability. Young's equation at point A then gives

$$\gamma(\cos\theta_0 - \cos\theta_{Ae}) = \gamma_1 \Phi_A. \tag{10.39}$$

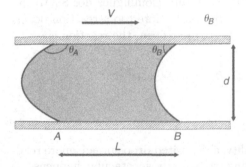

FIGURE 10.12. Column of a reactive fluid inside a capillary tube.

At point A, the reaction has proceeded for a duration $t = L/V$. To estimate Φ_A, we resort to a simple rate equation

$$\frac{d\Phi}{dt} = kc(1 - \Phi). \tag{10.40}$$

In this equation, k is a kinetic constant and c is the concentration of the reagent in the liquid. The factor $(1 - \Phi)$ describes the decrease in the number of available surface sites sustaining the reaction. For a surface continuously exposed during a time t, the solution is

$$\Phi_A = 1 - \exp(-t/\tau) \tag{10.41}$$

where the time constant τ is given by

$$\tau \equiv (kc)^{-1}. \tag{10.42}$$

Typically, the chemical reactions involved are relatively slow. The time constant τ is a few minutes at least. By contrast, the time t it takes for a drop to pass through is rather short (of the order of 0.1 s). Since $t \ll \tau$, we may simplify equation (10.41) to

$$\Phi_A \approx \frac{t}{\tau}. \tag{10.43}$$

Our description of the fundamental principles is now nearly complete. We write a relation (at point A) between force and velocity, analogous to equation (10.37):

$$F_A = \gamma(\cos\theta_{Ae} - \cos\theta_A) = \zeta_V V \tag{10.44}$$

where θ_A denotes the dynamical angle at the trailing edge.

This describes some dissipation near the contact line. But we have a much larger dissipation due to viscous losses in all the length of the liquid column. There is a difference $P_A - P_B$ between the two edges. In accordance with Laplace's law, we have

$$P_A - P_B = \frac{4\gamma}{d}(\cos\theta_B - \cos\theta_A). \tag{10.45}$$

This pressure difference sets up a Poiseuille flow with a velocity V such that

$$\frac{P_A - P_B}{L} = \frac{32\eta V}{d^2}. \tag{10.46}$$

At this stage, we introduce a simplifying approximation. In equation (10.45), which gives the pressure difference, we replace the dynamical angles θ_A and θ_B by their corresponding static equilibrium values θ_{Ae} and θ_0 (we will justify this approximation shortly). Under these conditions, equations (10.39) and (10.43) lead to

$$\frac{32\eta V}{d^2} = \frac{4\gamma_1 \Phi_A}{Ld} \approx \frac{4\gamma_1}{dV\tau}. \tag{10.47}$$

This expression gives a velocity V that is independent of the length L:

$$V = \left(\frac{\gamma_1}{\eta}\frac{d}{8\tau}\right)^{1/2}.\tag{10.48}$$

O. Rossier has studied the motion of reactive drops of octane containing perfluorinated trichlorosilane at a volume fraction level $\Phi_V = 0.01$. The capillaries were glass pipettes with an inner diameter of 1 mm. The capillaries were cleaned ultrasonically in a pure ethanol bath and subsequently dried in an argon flow. The quantities γ_1 and $\tau(c)$ were measured by monitoring the capillary rise of the reactive mixture. The results were $\gamma_1 = 2.8$ mN/m (corresponding to $V^* = \gamma/\eta = 5$ m/s) and $\tau(\Phi_V = 0.01) = 0.25$ s.

We arrive at drop velocities $V \approx (V^* d/8\tau)^{1/2} = 6$ cm/s. For $L = 0.25$ cm, the travel time t is 0.04 s, which, as announced earlier, is indeed much shorter than the time constant τ.

We now return to the approximations $\theta_A \to \theta_{Ae}$ and $\theta_B \to \theta_0$ in the pressure equation (10.45). For the sake of simplicity, we assume $\theta_0 \approx 1$ radian. The generic dynamical equation (10.44) gives angular deviations:

$$\Delta\theta = \theta_A - \theta_{Ae} \approx \theta_B - \theta_0 \approx \frac{\eta V}{\gamma}.\tag{10.49}$$

We proceed to compare these deviations to those produced by the chemical reaction, which are given by

$$\theta_0 - \theta_{Ae} \approx \frac{\gamma}{\gamma_1}\frac{t}{\tau} \approx \frac{t}{\tau}.\tag{10.50}$$

From there we deduce

$$\frac{\Delta\theta}{\theta_0 - \theta_{Ae}} = \frac{V}{V^*}\frac{t}{\tau}.\tag{10.51}$$

In the example considered above, $V \ll V^*$ and $t \ll \tau$, and we can be confident that our simplifying approximation is entirely justified.

10.3.3 Bidrops

Another way to generate a spontaneous movement is to create a train of dissimilar drops in a tube of diameter d.[16, 17] We will restrict ourselves to the case of *bidrops*, which consist of the juxtaposition of two drops of immiscible liquids (Figure 10.13).

FIGURE 10.13. A bidrop, resulting from the accretion of two drops of wetting and immiscible liquids. Such as system has the ability to move spontaneously.

If the two liquids are wetting, the bidrop system, which forms something like a two-car train, moves spontaneously because of its asymmetry! As the train moves, a bifilm is deposited in its wake, with two new interfaces being created (the first between liquid 1 and liquid 2, and the second between liquid 2 and air), while the film ahead is eliminated (the interface between liquid 1 and air). We deduce the driving force

$$F = \pi d(\gamma_1 - \gamma_2 - \gamma_{12}). \tag{10.52}$$

A movement will take place toward the left if liquid 2 spreads on top of liquid 1, in which case the spreading parameter S_{21} is such that $S_{21} = \gamma_1 - \gamma_2 - \gamma_{12} > 0.$*

> **Problem:** In the case when the liquids are only partially wetting, three menisci then merge with the tube with their respective angles θ_1, θ_2, and θ_{12}. Using Young's relation, show that $F = 0$. (Note: The same result can be arrived at with a surface energy argument. A displacement of the system at very low velocity does not deposit any film since wetting is only partial. There is, therefore, no energy gain resulting from such a displacement.)

The velocity of a bidrop depends on the constituent liquids. In the case when both liquids have the same viscosity η, and when that viscosity is high enough for inertia to be negligible, a straight equilibrium between F and a Poiseuille friction force [equation (10.47)] yields a constant velocity given by

$$V = \frac{1}{8} \frac{\Delta\gamma \cdot d}{L\eta} \tag{10.53}$$

where $\Delta\gamma = (\gamma_1 - \gamma_{12} - \gamma_2)$ and L is the length of the train. The length L is assumed to remain constant (as we have seen, the train does leak slightly, but the resultant change in dimension is generally quite negligible). In practice, V is of the order of a centimeter per second.

A more interesting case is when the viscosity of the "coal car" is very much larger than that of the "locomotive" ($\eta_2 \gg \eta_1$). In this case, the velocity is much greater than predicted by a Poiseuille law of the type embodied of equation (10.53). The reason is that the "coal car" can glide smoothly on the film of thickness e left behind by the "locomotive," and the friction force comes primarily from the velocity gradient V/e within the film. Dimensionally, the balance sheet of the forces involved can be written as $\eta_1 L_2 \, dV/e \approx \Delta\gamma \, d$, where L_2 is the length of the "coal car" and e obeys Bretherton's law [equation (5.37)]. Working through the derivation leads to $V \propto (\Delta\gamma \cdot d)^3/(L_2^3 \gamma_1^2 \eta_1)$, which indeed is independent of η_2. The

---

*Conversely, if liquid 1 spreads on top of liquid 2 ($S_{12} = \gamma_2 - \gamma_1 - \gamma_{12} > 0$), the bifilm is to the left and the drop will move toward the right under the action of a force $F = \pi d S_{12}$.

conclusion is that a bidrop has the ability to pull an ultra-viscous species in a confined environment.

10.3.4 "Running Drops" on a Solid Planar Surface

Principles

We now return to the case of reactive wetting. The capillary example described above is relatively straightforward to analyze. Unfortunately, it is quite a challenge to obtain "good," clean surfaces (free of hysteresis) inside a capillary tube. That is the reason why running drops were first observed and studied on planar, horizontal surfaces.

The ingredients remain the same, although the hydrodynamics is slightly more complicated to handle.[18] We can use the same framework as developed while analyzing the behavior of drops in a wettability gradient, the only difference being that here the drop itself imprints its own gradient S':

$$S' = \frac{\gamma_1 \Phi}{l} \approx \frac{\gamma}{l} \frac{t}{\tau} = \frac{\gamma_1}{V\tau}. \tag{10.54}$$

Using equation (10.16), we find

$$V = \frac{2h_0}{3\eta \ln} \frac{\gamma_1}{V\tau}. \tag{10.55}$$

From this we deduce

$$V^2 = \frac{\gamma_1}{\eta} \frac{\theta_0 l}{\tau} \cong V^* \frac{l}{\tau} \tag{10.56}$$

where we have assumed $\theta_0 \approx 1$ and $\gamma_1 \approx \gamma$. Here again, the velocity is a geometric average of a capillary velocity (V^*) and a chemical velocity (l/τ).

We expect the result to be $V \propto \sqrt{lc}$, where c is the surfactant concentration, which controls τ via equation (10.42). These predictions are reasonably well borne out by the experiments reported in reference 14.

Discussion

- How does the motion of the drop get started? The question has been studied by Shanahan, who examined the fluctuation modes of a drop at rest.[19] He found two main modes:

  1. A breathing mode, whereby the drop remains centered on a fixed point but changes its degree of spreading.
  2. A polar mode, whereby the center of gravity moves. This mode is unstable and gradually spawns a translational motion that ultimately reaches a constant velocity.

- A group at Oxford University did an entertaining experiment by tilting the supporting plate.[13] They observed some drops climb up the slope, slow down, stop, and eventually start on their way back down.

- Returning to a horizontal support, drops collide with the lateral walls surrounding the plate. They are then reflected and resume their motion, following complex trajectories. For better control of the experiment, it is sometimes advantageous to confine a drop between two "rails" (non-wettable ribbons).[14]
- In the absence of rails, a drop may attempt to traverse a prior trajectory. Such a region turns out to be "hostile" (less wettable), and the drop cannot always get across. The discussion of crossing events is complicated because inertia must be taken into account.
- Another fascinating question concerns the movement of drops on random surfaces (where the wettability varies from point to point). If a drop has a size comparable to that of the defects, it undergoes frequent collisions and quickly forgets its initial direction. If, on the other hand, the drop is large in comparison with both the size of the defects and the distance between them, one ought to observe a motion that remains straight over a fairly large "persistence length."[20] The difficulty is in creating forces $(\gamma_1 \Phi_A)$ large enough to overcome hysteresis.
- One might wonder whether spontaneous motion is a property specific to running *drops*. Could other objects, such as a *gel ball* containing a reactive solution, exhibit a similar behavior? For example, one could take a latex particle and inject into it an alkane/chlorosilane solution. Would such an object roll on a glass surface? If it did, it would qualify as a "running sponge."[21] It is difficult to predict the behavior of such sponges. The particle still has a "footprint," i.e., a finite contact area with the support. One must determine if the adhesion of the footprint is strong enough to prevent any motion. As the object rolls with a velocity V, one must also know whether the dissipation at the footprint can dramatically slow down the movement. The effect is certainly possible conceptually, but no one has observed it todate.

10.4   Transport by Electric Field

10.4.1   Relevance of Microsystems

Various practical problems demand that liquids be transferred from one tiny compartment to another. Three examples follow:

1. Analysis of a bacteria population in ambient air. The technique consists in first collecting the bacteria in a small dish, then cracking them open by means of an electric field, and finally transferring their DNA to a series of reactors, each of which contains a segment of DNA specific to a particular bacterium. If the DNA under test matches any of the references, recombination occurs, which can be observed by fluorescence methods. Everything must be done on a small scale to limit the quantities of material needed and the analysis time.

2. Altering the turbulent flows near the wings of supersonic aircraft. The objective is to tailor airflows in the turbulent boundary layer using small protruding pins that control the birth of vortices.

3. Ejection of drops at high velocities in high-speed printers. To promote high speed, one typically resorts to very small structures. Such structures can be etched in silicon by standard techniques. However, the resulting systems must be active in the sense that they are capable of displacing fluids. Several methods exist:

- by creating a hot spot in a microresistor and generating a bubble, which, in turn, displaces a fluid and can also serve as a barrier between two regions,
- by attracting a dielectric fluid in the field between the plates of a microcapacitor,
- by deforming one of the walls of a container by means of a membrane.

In the present section, we will focus on somewhat different electrical effects, the practical applications of which remain perhaps limited so far, but which are directly related to capillary effects—electrocapillarity[22] and electro-osmosis.

10.4.2 Electrocapillarity

The surface tension γ_{12} of the mercury/water system changes as one applies a voltage U to mercury.

The functional dependence of $\gamma_{12}(U)$ depends critically on the type of ions present in the solution. Suppose the solution contains sodium sulfate. In such a case, in the absence of an applied voltage, the sodium ions Na^+ attach themselves to the mercury, and a double layer forms in the water with nothing but SO_4^{2-} ions over a certain screening length, roughly 30 Å in size. The double layer acts like a capacitor. With no externally applied bias, the capacitor charges spontaneously by adsorption. If a positive bias is applied, the Na^+ ions begin to be expelled, and for a particular bias U_p, one recovers the surface tension γ_{12}^0 of pure water. We may then write an idealized functional dependence:

$$\gamma_{12}(U) = \gamma_{12}^0 - \frac{1}{2}C(U - U_0)^2 \tag{10.57}$$

where C is the capacitance per unit area of the double layer. The second term represents the decrease of energy due to polarization.

If the surface tension depends on the applied bias, it becomes possible to set a mercury column in motion in a capillary tube by applying different voltages to either end. Hence the possibility of fabricating micropumps on a tiny dimensional scale.

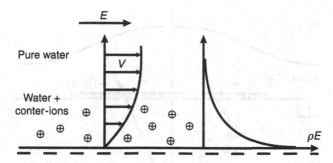

FIGURE 10.14. Principle of electro-osmosis. The local force acting on the liquid (produced by an electric field E acting on a charge density ρ) creates a flow.

10.4.3 Principle of Electro-Osmosis

An understanding of electro-osmosis goes back to Schmoluchowski (Figure 10.14). An electrically charged surface (surface potential ζ) exposed to a layer of almost pure water develops a double layer (surface + counter-ions), the thickness of which is the screening radius where the counter-ions collect.

Under the influence of a tangential electric field E, these ions start moving and induce within the water film a plug flow with a velocity given by

$$v = -\frac{\varepsilon E \zeta}{\eta} \tag{10.58}$$

where η is the viscosity and ε is the dielectric constant.

We consider here evaporated electrodes separated by a distance of ~ 100 µm and subjected to a potential difference of $\pm\ 1$ V. It follows that the electric field is ~ 100 V/cm. With typical values of ζ_E of 100 mV, we get $v \approx 0.5$ mm/s. In other words, water moves by one unit (100 µm) in 0.2 s.

Two possible applications are worth mentioning:

- *Modifying wetting phenomena.* For example, a drop deposited under partial wetting conditions must be set in motion. If the only important effect is that described in equation (10.58), the drop will move without deformation. On the contrary, if electrolysis effects alter the support, the drop will not have the same equilibrium angle at the leading and trailing edges. Such effects can be avoided with the proper choice of inert electrodes (gold, graphite, etc.). Electro-osmosis may possibly inhibit dewetting (in partial wetting) or even arrest a precursor film (in total wetting).

10.4.4 Examples

10.4.4.1 Electrostatic Lenses

Electric fields E are created at the outskirts of a conducting plate (Figure 10.15). In the presence of positively charged counter-ions, a water film

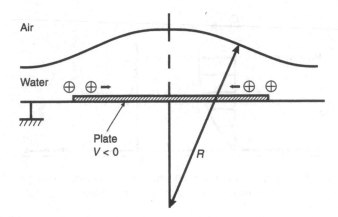

FIGURE 10.15. Electrostatic lens created by a radial electric field.

will tend to thicken near the central point. A pressure field $p(x, y)$ develops to balance out the effect of electro-osmosis. The resulting flow v is given by

$$v = -\frac{\varepsilon E \zeta}{\eta} - \frac{h^2}{3\eta} \nabla p. \qquad (10.59)$$

At equilibrium, we have $v = 0$. The pressure $p(x, y)$ and the electrical potential $U(x, y)$ are proportional to each other:

$$-\frac{\varepsilon \zeta U(x, y)}{\eta} = \frac{h^2}{3\eta} p(x, y). \qquad (10.60)$$

The pressure is related to the radius of curvature R of the lens through Laplace's equation:

$$p = \frac{2\gamma}{R}. \qquad (10.61)$$

It follows that

$$\frac{1}{R} = \frac{3}{2} \frac{e\zeta}{\gamma h^2} U(x, y). \qquad (10.62)$$

For $h = 10$ μm, $\zeta = 100$ mV, $U = 1$ volt, and $\gamma = 70$ mN/m, the above equation yields $R \approx 20$ cm.

Question: If the bias is removed, how long does it take for the lens to disappear?

Answer: The capillary modes of a viscous film are of the form (discussed in chapter 5)

$$\omega(q) \propto \frac{\gamma}{\eta} q^4 h^3. \qquad (10.63)$$

Here $q \cong \pi/l$, where $l = 100$ μm is the lens diameter. We deduce that $\pi/\omega \approx 10^{-4}$ s.

FIGURE 10.16. Transfer of a bubble by electro-osmosis.

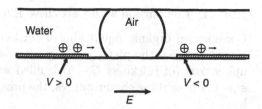

10.4.4.2   Transfer of Bubbles

The principle is sketched in Figure 10.16. Here, a liquid film of thickness h is trapped between two solid, insulating plates, one of which has contact pads. The liquid contains a bubble.

When applying a bias between the two pads, the bubble is set in motion at a velocity v [equation (10.58)] and crosses from one pad to the other in a time $\tau = l/v = 0.05$ s. We thus have a bistable system, which can be switched on and off, with good optical contrast.

10.4.4.3   Limitations

Thermal Limitations

If the field E is applied continuously, there is significant Joule heating of the water, which can be a serious drawback.

- For relatively pure water (Debye length $\kappa_D^{-1} = 20$ Å) in an electric field $E = 100$ V/cm, the temperature rise is estimated to be 0.10 K per minute under the worst conditions (when the support is not cooled). This means that we can afford to leave the field on for a few minutes only. In a pulsed format (pulses shorter than ~ 1 s), the electric field can be as high as 10^3 V/cm.

- In bioengineering, it is often necessary to work with salt solutions, in which case water conducts more readily and the thermal limitations are even more severe.

Electrochemical Limitations

If there is substantial release of gas or oxidation of the pads, the device characteristics are likely to degrade rapidly. We may be forced to resort to noble metals (such as gold or platinum) in thin layers, or to carbon, or even to semiconductors (such as silicon). In any event, it is necessary to keep the applied bias below the threshold for electrolysis of water (~ 1 V).

Contamination by Air

We may want to protect the film of water by covering it with some organic liquid film.

10.4.4.4 Comparison with Capacitive Effects

Consider an organic liquid film of thickness h and dielectric constant ε placed between the plates of a capacitor separated by a distance D. The upper part (of thickness $D - h$) is filled with air. We assume $h \ll D$ and $\varepsilon \gg 1$. Under these circumstances, the pressure induced in the film is given by

$$p_{el} = \frac{-\varepsilon_0 E_0^2}{2} \qquad (10.64)$$

where $E_0 = U/D$, U being the voltage applied between the two plates. This pressure is negative wherever $E_0 \neq 0$, which means that the liquid is attracted toward the regions of strong field. In other words, by applying localized fields E_0, it is possible to transport fluids, move bubbles, form lenses, etc.

These capacitive effects tend to be small because the field E_0 is weak on an atomic scale and the pressure varies as the square of the electric field. Nonetheless, larger field values can be used with capacitive effects than with electroosmosis, since we are dealing here with insulating liquids. Here again, we can create lenses with a localized field. The main advantage of these organic liquids is to be electrochemically inactive. Referring to Figure 10.15, if we chose to work not with a liquid/air system, but with two organic liquids with different dielectric constants ε, the entire system could be sealed in an enclosure and we might hope that chemical changes might not take place for very large numbers of cycles.

Conclusion. We may expect transport phenomena by means of electric fields to spawn a plethora of new devices. Electro-osmosis enables us to select specific flow directions, paving the way toward intriguing experiments. However, as far as practical technological applications are concerned, capacitive effects hold great promise because of their superior robustness.

References

[1] P. Carles and A. M. Cazabat, *Colloids & Surfaces* **41**, 97 (1989).

[2] T. Ondarçuhu and E. Raphaël, *C. R. Acad. Sci. (Paris)* **314**, 453 (1992).

[3] M. Chaudhury and G. M. Whitesides, *Science* **256**, 1539 (1993).

[4] F. Brochard, *Langmuir* **5**, 432 (1989).

[5] J. B. Brzoska, F. Brochard, and F. Rondelez, *Langmuir* **9**, 2220 (1993).

[6] J. B. Brzoska, F. Brochard, and F. Rondelez, *Europhys. Lett.* **19**, 97 (1992).

[7] S. Troian, E. Herbolzheimer, S. A. Safran, and J. F. Joanny, *Europhys. Lett.* **10**, 25 (1989).

[8] A. M. Cabazat, F. Heslot, S. Troian, and P. Carles, *Nature* **346**, 824 (1990).

[9] H. E. Huppert, *Nature* **300**, 427 (1982); N. Silis and E. B. Dussan, *Phys. Fluids* **28**, 5 (1985).

[10] F. Melo, J. F. Joanny, and S. Fauve, *Phys. Rev. Lett.* **63**, 1953 (1989).

[11] C. Redon, F. Brochard, and F. Rondelez, *J. Phys. II France* **2**, 1671 (1992).

[12] W. Zisman, Contact Angle, Wettability and Adhesion, *Adv. In Chem. Series* **43**, 1 (Washington, D.C.: American Chemical Society, 1964).

[13] C. Bain, G. Burnett-Hall, and R. Montgomerie, *Nature* **372**, 414 (1994).

[14] F. Dominguez dos Santos and T. Ondarçuhu, *Phys. Rev. Lett.* **75**, 2972 (1995).

[15] J. T. Davies and E. K. Rideal, *Interfacial Phenomena* (New York: Academic Press, 1963), chapter 6.

[16] C. Marangoni, *Ann. Phys. Chem.* **143**, 337 (1871).

[17] J. Bico and D. Quéré, *Europhys. Lett* **51**, 546 (2000); J. Bico and D. Quéré, *J. Fluid Mech.*, **467**, 101 (2002).

[18] F. Brochard and P. G. de Gennes, *C. R. Acad. Sci. (Paris)* **321 II**, 285 (1995).

[19] M. Shanahan and P. G. de Gennes, *C. R. Acad. Sci. (Paris)* **324 II b**, 261 (1997).

[20] P. G. de Gennes, *C. R. Acad. Sci. (Paris)* **327 II b**, 147 (1999).

[21] P. G. de Gennes, *C. R. Acad. Sci. (Paris)* **323 II**, 633 (1996).

[22] C. Quilliet and B. Berge, *Current Opinion in Colloid and Interface Science* **6**, 34 (2001).

Index